Used tyres in solid waste management

ORGANISATION
FOR ECONOMIC
CO-OPERATION
AND DEVELOPMENT
PARIS 1980

The Organisation for Economic Co-operation and Development (OECD) was set up under a Convention signed in Paris on 14th December 1960, which provides that the OECD shall promote policies designed:

— to achieve the highest sustainable economic growth and employment and a rising standard of living in Member countries, while maintaining financial stability, and thus to contribute to the development of the world economy;
— to contribute to sound economic expansion in Member as well as non-member countries in the process of economic development;
— to contribute to the expansion of world trade on a multilateral, non-discriminatory basis in accordance with international obligations.

The Members of OECD are Australia, Austria, Belgium, Canada, Denmark, Finland, France, the Federal Republic of Germany, Greece, Iceland, Ireland, Italy, Japan, Luxembourg, the Netherlands, New Zealand, Norway, Portugal, Spain, Sweden, Switzerland, Turkey, the United Kingdom and the United States.

Publié en français sous le titre:

LES PNEUS USÉS
DANS LA GESTION DES DÉCHETS

***

D
628.54
ORG

© OECD, 1981
Queries concerning permissions or translation rights should be addressed to:
Director of Information, OECD
2, rue André-Pascal, 75775 PARIS CEDEX 16, France.

CONTENTS

SUMMARY AND CONCLUSIONS

As a result of the very large increase in car ownership during the last two or three decades there has been a dramatic increase in the number of used tyres to be disposed of. In the United States, over two hundred millions tyres are scrapped each year, that is, about one tyre for every person in the country. This relationship holds roughly in the other developed countries. Since tyre production is highly energy intensive, this means that every year billions of gallons of fuel oil equivalent are being discarded through tyre disposal.

In the total solid waste stream of OECD Member countries used tyres represent only a small proportion (0.5 to 1.0 percent) but their physical durability and chemical composition give rise to major environmental problems. Because they are designed to resist most forms of degradation they create severe handling and operational difficulties at disposal sites. Incineration of used tyres generates large quantities of unsightly and noxious smoke, polluting the atmosphere, and their high thermal content can damage the incinerators.

In 1976, the OECD Council adopted a Recommendation urging Member governments to develop and implement comprehensive waste management policies which would meet both environmental protection and economic objectives, and instructed the Environment Committee to elaborate the main elements of such policies and to facilitate their practical implementation by promoting cooperation among Member countries. In this Recommendation, a comprehensive waste management policy is defined as "a coherent system of measures concerning the design, manufacture and use of products as well as the reclamation and disposal of waste, and aiming at the most efficient and economic reduction of the nuisances and costs generated by waste". As far as automobile tyres are concerned the main elements of a comprehensive waste management policy may comprise the following options: the reduction of waste generation at source by making longer-lasting tyres; the re-use of chemically unaltered material, in whole tyre or other form, in particular through retreading; the reclaiming of waste rubber to recycle it in the manufacture of new rubber products; and the recovery of heat value, constituent materials or chemicals from discarded tyres.

Tyre life has increased over the years, particularly since the development of the steel-belted radial ply tyre. The longer-life tyres represent a higher proportion of the market for new cars than for replacement purposes, however. Retreaded tyres are taking a small proportion of the replacement market for private cars, although the share is higher for trucks, buses and similar vehicles. Aeroplane tyres are retreaded many times over.

Traditional waste management options for tyres include: landfill, the production of rubber crumb by grinding, the chemical production of 'reclaim', and retreading. The advent of the steel reinforced tyre has reduced the production of rubber crumb; also the production costs and the market price of recovered materials have significantly reduced the production of reclaim from discarded tyres.

More recent alternatives for management include the cryogenic crushing of tyres for the production of crumb (enabling steel belted radials to be ground); the use of rubber crumb for various road-construction purposes: the use of crumb in other indoor surfaces; the pyrolysis of tyres; and tyre incineration. Each option is evaluated in the text. None appears to have general application (with the possible exception of crumb for roads in Canada) but each seems to be feasible in particular regional or local applications.

The encouragement of increased retreading as a management alternative for used tyres requires that programmes be implemented to ensure adequate quality control. The production of longer life (100,000 miles or 160,000 kilometers) tyres has been shown to be feasible in the United States, and bears investigation for other countries.

The study has examined all the options currently available to manage the scrap tyres generated each year, and to reduce this volume. On the information available, the evidence seems to indicate that:

i) all currently-available disposal options have net external costs associated with them;

ii) of these disposal options, the one which presents most flexibility for waste tyre management is the storage of shredded scrap tyres. This policy would need to be accompanied by research and development into the more efficient recovery of chemical constituents from used tyres, in anticipation of further significant rises in the price of petrochemical feedstocks into the tyre industry;

iii) retreading is commercially viable as well as socially desirable, and fiscal and regulatory measures to increase the volume of car tyres retreaded should be considered;

iv) extension of tyre life to 100,000 miles yields high social benefits, and the tyre industry should be encouraged to produce such tyres on a large scale as soon as possible.

Reducing the volume of scrap tyres, either by source reduction or through resource recovery, can be encouraged by a number of measures including regulatory requirements and economic incentives. In particular, the feasibility of imposing a 'product charge' on tyres should be thoroughly examined.

INTRODUCTION

Used tyres represent only a small fraction of the total solid waste stream arising in Member Countries. The management of this category of solid waste, whilst difficult, is however particularly necessary. The difficulties represented by used tyres in the solid waste stream stem both from the physical characteristics of the tyre, and from its chemical constituents. These present significant environmental problems if either of the conventional disposal options is used. If whole tyres are incorporated in bulk into a sanitary landfill site, the following problems can arise:

i) they can flex back to their original shape after compacting, and work their way to the surface while the fill is settling

ii) if left on the surface they can harbour disease carriers such as rats

iii) tyres in landfill represent a significant fire hazard

iv) tyres are resistant to natural decomposition, and thus represent a long-lived and cumulative solid waste problem.

The other conventional solid waste disposal option, incineration, also has several adverse impacts:

i) tyre incineration generates large quantities of unburned hydrocarbons, producing highly unsightly and noxius smoke. This can only be controlled by expensive retrofitted equipment, including an augmented air supply followed by scrubbers or electrostatic precipitators

ii) whole tyres have such a concentrated high thermal content that burning them in modern incinerators can damage the refractories.

This complex set of environmental impact problems that waste tyres present has resulted in a great deal of work in Member Countries on various aspects of the problem. This report is based upon this work, augmented with work from other countries where relevant.

This report is structured as follows: Chapter One summarises the development of tyre technology to the present day, and mentions a few developments which may affect the mix of tyres entering the waste stream in future years. Chapter Two discusses the tyre cycle, tracing the major paths through which used tyres are generated, and the major disposal or resource recovery routes followed. Chapter Three examines data from Member countries on the proportion of the market for original equipment and replacement tyres taken by the two major types of tyre, radial and cross-ply. Chapter Four examines current waste management options used in Member Countries, and the main thrust of innovative research on waste tyre management options in these countries. Chapters Five to Eight evaluate the major alternatives for used tyre management, namely physical, chemical and energy recovering alternatives, and the options of increasing tyre life either by more retreading of tyres, or by making even longer-lived original equipment tyres. The final chapter summarises the discussion of the previous chapters, and draws out policy recommendations that emanate from them.

Chapter I

TYRE TECHNOLOGY

Over recent years, developments in tyre technology have resulted both in changes in tyre construction methods, and in the reinforcing materials used in tyres. The tendency has been to produce tyres which, with ideal maintenance and careful use, will last longer, but which, when discarded, are typically less suitable for retreading, and from which the rubber is more difficult to reclaim. Hence, used tyre management poses an interesting question as to the relative social merits of these two solid waste management options.

The most widely used tyre for many years has been the "cross-ply" tyre. This consisted of textile impregnated with a rubber compound, layers of which, with the cord (textile material) running diagonally (alternate diagonals for alternate layers) formed the basis of the tyre. The tyre has steel bead wires included at the extremes of the reinforcing plies (cord) to ensure a complete seal onto the rim. (1) The outer surface of the tyre casing is covered with rubber compounds. That used for the walls is flexible. That for the tread is abrasion resistant.

The radial ply tyre is now replacing the cross-ply tyre as the predominant tyre used in many countries. In this tyre construction, reinforcing plies are wound radially around the tyre (at right angles to the direction of rotation). This gives the tyre greater flexibility. Further technical developments have included the use of alternative reinforcing materials, such as nylons, terylenes, steel and fibreglass.

More recently, some new concepts in tyre design include one now on the market and one in the final stages of development. The former has "run-flat" capabilities. It is a steelbelted radial tyre that can be built up on existing equipment, using similar rubber compounds to those currently used. However, the design of the tyre, and the provision of lubricants inside the tyre minimise friction when tyre is deflated, and enable it to be run for more than 80 kilometres (50 miles) at speeds of up to 80 k.p.h. (50 m.p.h.) without risk of tyre damage, in the event of accidental deflation whilst running. This tyre is now used, either

1) The bead wire is made of many turns of small diameter wire wrapped in a rubber impregnated tape. This forms a flexible hoop of high tensile strength.

as original equipment or as an option, on cars in Great Britain and North America. Its widespread adoption would have far-reaching implications for used tyre management. The latter new tyre concept is the 'compression sidewall' tyre, which also has a 'run-flat' capability.

These trends in tyre technology have obvious implications for the recycling/product life extension ("PLE") trade-off. Most of the new tyre developments involve the increased use of steel reinforcement in the tyre belt. The presence of this steel makes the tyre increasingly difficult to recycle, but also has the effect of increasing the life of the tyre under appropriate conditions of use. The returns to further prolongation of tyre life, versus investment into research and development for reclaiming the rubber compounds from tyres with a higher steel content, needs careful consideration.

Several European countries, and Australia, all report that radial ply tyres are taking an increasing share of the tyre market, with an increasing proportion of these being steelbelted radial ply tyres. In North America, the trend is somewhat different. The more firm ride characteristics of the steelbelted radial ply tyre are not so suitable for the North American market, but the fibreglass reinforced radial tyre has made progress there. Annex 1 contains information on the penetration of radial ply tyres in the original equipment and replacement tyre markets in several Member Countries.

One other tyre technology that has recently reemerged after having been rejected in the 1960's is worthy of mention here [7]. The technology is essentially that of producing a tyre by injection moulding, and was originally developed by the Caterpillar Corporation of the United States for use on earthmoving equipment. A scaled-down prototype, suitable for automobiles, was developed, but was rejected by the three major automobile companies. Possible reasons for this include loss of market share (as the tyre was expected to last 100,000 miles or more) ; possible labour trouble, as the process was essentially one of low skill, mass production (in contrast to the conventional tyre building process, which is both labour intensive and highly skilled) ; and the fact that the major United States tyre companies were, at this time, undertaking a massive investment programme for the production of steelbelted radial ply tyres. It has been suggested that the tyre companies wished to recoup their return on this investment before undertaking another major investment programme. (1)

---

1) There is also information that a major European tyre manufacturer has also recently produced and tested a tyre by injection moulding, but that the tyre's appearance on the market will be delayed for similar reasons.

Chapter II

THE TYRE CIRCLE

A thorough evaluation of waste tyre management policies requires a knowledge of the ways in which waste tyres arise: the path from manufacturer through use to discard, and the existing waste management and resource recovery options. Particular information on individual countries is contained in Annex II. The following section is a generalised version of the tyre cycle, from which the actual cycle in particular countries may diverge with respect to institutional details.

Figure 1 shows the generalised tyre cycle. From the manufacturer tyres are brought into use through an extensive distribution network to tyre stockists, who may be integrated with the manufacturer, or may be independent. The stockists (which range from garages and filling stations selling tyres for private vehicles through to specialist tyre stockists) often specialise in one or two of the five major categories of use (cars and motorcycles, trucks, buses, aviation, and construction and agriculture). Aeroplane tyres are a specialist trade, and tend to be handled separately from other categories, the companies dealing with them often being integrated into the aviation industry.

When the initial tread is worn down to the minimum acceptable standard, or when carcass damage prevents the tyre being used safely, the tyre enters the inventory of used tyres. These tyres may then be sent to landfill, or be chosen as suitable for retreading. Tyres may be shredded or ground to a crumb. Other possible uses for used tyres, based on a resource recovery rather than waste disposal philosophy, are the use of rubber crumb in roads, or the pyrolysis of tyres to recover the gaseous, oil and char fractions (which may be used in tyre manufacture). Thus, the tyre cycle is, to some extent, a closed loop. If reclaim rubber, or pyrolysis fractions, are used in tyre manufacture, or if tyres are retreaded, then this is the case. If tyres are sent to landfill, then useful materials will be lost. Energy recovery represents an intermediate stage, where the energy component of the tyres is recovered by incineration. Such an approach may however create as many environmental problems as it solves.

The structure of the tyre industry varies slightly from country to country, and so, consequently, do the routes by which tyres are

Figure 1 **THE TYRE CYCLE**

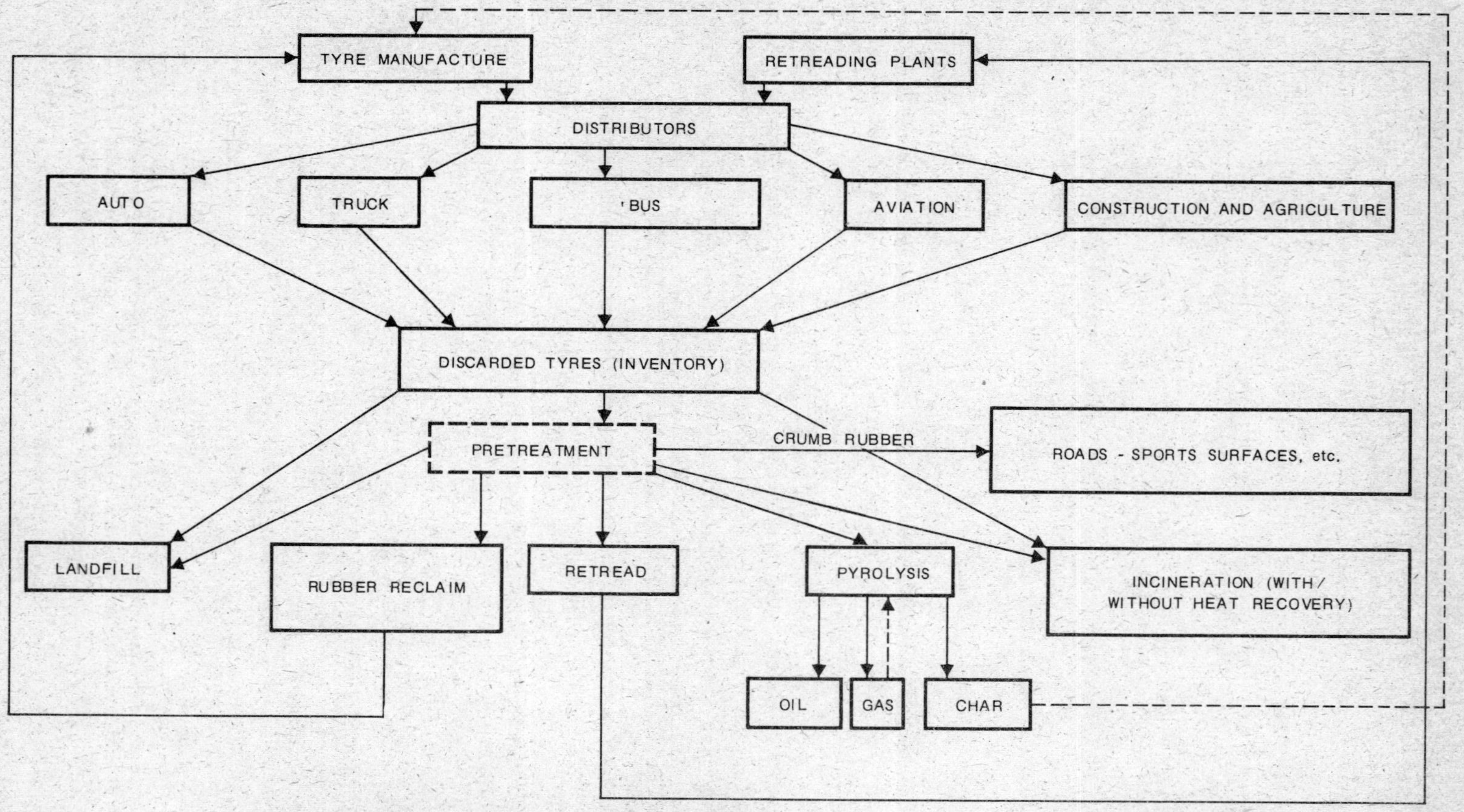

channeled to ultimate disposal. One of the key features of the "tyre cycle" is, however, the fact that most used tyres are accumulated by specialist tyre retailers, independent filling stations who retail tyres, and by vehicle dismantlers. In this respect, tyres differ from general municipal waste, which has to be collected from each individual household.

Chapter III

WASTE TYRE ARISINGS

1. REASONS FOR DISCARDING TYRES

Whilst the sales of new tyres are relatively easy to estimate, used tyre arisings are difficult to predict. The life of a tyre is determined by several factors:

- construction (radial or bias-ply, etc.)
- inflation
- reinforcing material used (fabric, fibreglass, steel)
- type of driving
- type of road surface
- quality of maintenance of vehicle (especially shock absorbers)
- ambient atmospheric temperature.

Of these, inflation, construction, and reinforcing material are thought to have the most significant effects. In Canada, it is estimated that, on average, radial ply tyres last for 56,000 kms (35,000 miles), belted bias tyres for 40,000 kms (25,000 miles) and bias ply tyres for 24,000 kms (15,000 miles) [23, p. 83].

One relationship between driving style and tyre wear is shown in table 1. The faster the rate of cornering, the higher the rate of tyre wear (with the rate of increase increasing with the speed). Qualitatively similar results hold for braking and acceleration.

Table 1

RELATIONSHIP BETWEEN DRIVING SEVERITY AND TYRE WEAR *

Speed on curves (k.p.h.)	Specific wear rate (mm/1000 km)
32	0.13
45	0.50
57	1.10
67	3.10

* Source: 6, p. 21.

Table 2 shows one U.S. manufacturer's estimate of the effects of different degrees of underinflation on tyre life.

Table 2 *

EFFECT OF INFLATION ON TYRE LIFE IN ACCORD WITH A MAJOR MANUFACTURER'S EQUATION **

Inflation (PSI)	Percent of life at 24 pounds per square inch.
8	45.5
10	54.4
12	62.5
14	70.1
16	77.0
18	83.3
20	89.4
22	94.8
24	100.0

*) Source: 9, p. 18d.

**) Percentage of life at 24 PSI = $\frac{\text{Inflation}}{[0.401\ (\text{Inflation}) + 14.338]}$

From these figures it appears that one major policy option open to Member countries to reduce arisings of waste tyres is to provide more incentives for proper tyre maintenance, and for more careful driving. This will be discussed in greater detail below.

A German study [2, pp. IV/15-18] has identified the following factors as important determinants in the average life of a tyre:

- average vehicle mileage per year
- average traffic density
- average velocity
- engine capacity of the vehicle
- the quality of the road surface
- proportion of winter tyres in total tyre sales
- proportion of retreaded tyres in the total tyre population
- minimum tread depth allowed on tyres

The factors are mentioned only qualitatively. There seems little chance to quantify the separate effects of each of these factors on average life in any systematic fashion.

2. NUMBERS OF WASTE TYRES GENERATED

Methods of estimating used tyre arising vary greatly. In Australia, it is estimated that, with existing tyre technology, approximately 0.7 tyres are discarded per person per year. Although this is a rule of thumb, its predictive usefulness is quite significant. As well as the above assumptions, however, the predictive usefulness of this rule of thumb depends also upon the proportion of retreaded tyres remaining constant in total tyre sales. Using this approach, 6,757,000 tyres were scrapped in cities, town or regions with populations of 30,000, which covers 75 % of the Australian populations. Using this estimate, approximately 9,009,000 tyres are scrapped in Australia each year.

In the United Kingdom, the methodology for estimating used tyre arisings is rather different, although the Australian procedure must implicitly have a similar base. Using this system, scrap tyre arisings can be estimated separately for private cars and for trucks and buses. These are shown in Annex 3.

The accounting system used to calculate the arisings of used tyres in the United Kingdom is of real interest, as it is probably the most thorough system used in Member countries to evaluate the volume of used tyres generated [21, pp. 90-91].

The system uses the following set of equations and identifies:

Home arisings (Used tyres) = Home replacement sales (new tyres + retreads) + all tyres from scrap vehicles (1)

by definition:

Home arisings (used tyres) = Home arisings (scrap tyres) + home arisings (tyres for retreading) (2)

Thus:

Home arising (scrap tyres) + home arisings (tyres for retreading) = Home replacement sales (new tyres + retreads) + all tyres from scrap vehicles (3)

But (assuming no change in retreaders inventories):

Home arisings (tyres for retreading) = Home replacement sales (retreads) (4)

Therefore:

Home arisings (scrap tyres) = Home replacement sales (new tyres) + all tyres from scrap vehicles (5)

Several points about this approach need to be noted:

i) Home replacement sales include imports for home replacement purposes.

ii) Equation (4) is only an approximation: the number of tyres going to retreading is only approximately equal to the number of retreaded tyres going to the replacement market.

iii) When arisings of waste tyres need to be separated into proportions of radial and cross-ply tyres, due to substitution of radial ply tyres for cross-ply tyres, care has to be exercised, and the procedure modifed slightly.

The volume of used tyres estimated according to this model is shown in Annex 3. Another approach to estimating the volume of used tyres is one used in a United States study, where magnitudes are placed on the various flows in the relevant tyre cycle /9, p. 6a/.

The situation in Japan is reported in a recent study /18/. In 1976, it was estimated that about 42 million tyres were replaced in that year alone. This represented 55 % of the total number of tyres sold in Japan. During the years 1974-1976, it was estimated that, of the tyres removed from vehicles, 75 % came from the replacement of worn tyres by new tyres, and 25 % came from scrapped vehicles. These 75 % were accumulated at the premises of those enterprises selling tyres. A survey in 1974 showed that the main enterprises involved were: tyre sales agencies; speciality tyre stores; and (a relatively small number of) waste tyre dealers /18, p. 4/

3. THE MANAGEMENT OF USED TYRES

Given that the volume of discarded tyres is increasing over time, and is projected to increase through the next decade, unless disposal facilities are increasing proportionately, the problem of waste tyre disposal will become increasingly serious. The limited evidence available suggests that the conventional disposal routes (landfill, simple incineration, reclamation and recovery) are becoming increasingly less able to cope with the volume of waste tyres being generated. There are several reasons for this:

i) for many metropolitan areas, landfill sites are becoming increasingly scarce, and transport costs to these sites are increasing. In addition, the safety and health hazard problems associated with waste tyre disposal have led to some sites refusing to accept whole tyres

ii) higher speed and longer distance motoring have led to specifications for, and governmental regulation of, tyres being made more stringent. This, in turn, has had two effects: the retreaded tyre has been thought to be less suitable for today's driving conditions and the compounds reclaimed from tyres to be less suitable for inclusion in the fabrication of higher speed, higher performance tyres

iii) tyres that are driven for increasingly longer distances under more severe conditions are, when the tread is below its minimal safe level, less suitable for retreading, due to greater tyre wall damage, for exemple.

It is important to remember that, even with the tread worn down below its minimal safe depth, a tyre still has approximately 90 % of its original material content. Although natural rubber comes from a renewable resource (mainly Hevea brasiliensis) many of the elastomers used in tyre manufacture are synthetic, derived largely from petrochemical feedstocks. Thus, the opportunity cost to a society of conventional disposal, either through landfill or through simple incineration without heat reclamation, is likely to be high. However, high transport and selection costs militate against retreading. The grade of rubber that can be produced mechanically from used tyres has a limited demand, and the move towards steel reinforcing has reduced the proportion of tyres that can be handled in this way. More recently-developed methods of reclamation, including chemical separation, cryogenic separation and pyrolisis, either produce a product of limited market value, or are extremely costly to run (or both). In all cases, the high costs of the process are compounded by the transport costs involved in assembling the tyres at the point of reclamation. Special furnaces are necessary to ensure complete combustion of the tyres, and to control the gases and ash produced. The alternative is to encourage more satisfactory tyre maintenance, and to encourage the manufacture of tyres with a significantly longer useful life in its primary use. It is to be expected that the optimal management policy for used tyres will involve used tyres going via the various disposal routes, and a further reduction in waste arisings being achieved by tyre life prolongation. The relative merits of each alternative are discussed below, together with some cost estimates for various processes.

As with most forms of solid waste, collection and transport costs constitute a significant proportion of the total cost of any recycling or disposal scheme. Particularly in countries with large land areas, collection and transport costs can determine whether disposal and reprocessing facilities should be on a large scale and centralised, if there are scale advantages to this; or, whether such facilities should be localised, to reduce transportation costs. This problem presents itself acutely in Australia, for example [6, p. 24]. In some countries, tyres accumulated on tyre dealers premises are transported back to regional depots as back loading. The extra costs involved are so small that these "initial concentration" costs are ignored. Given the coarseness of much of the data on solid waste management in general, such a procedure is difficult to question. In Great Britain, the situation is similar to that in Australia. Tyre manufacturers deliver new tyres to their depots, who sell directly to customers and also supply retail outlets. A manufacturer is supposed to collect all worn tyre casings from depots and garages, taking them back to the factory for sorting. There is supposed to be a "casing bank", so that for each new or remoulded tyre supplied to an outlet, a used tyre is returned. In practice, there are specialist buyers for good quality used tyres, resulting in the manufacturer only receiving the casings less suited or unsuited to retreading or remoulding. If an outlet persists in this practice, collections may be suspended by the manufacturer. The "tyre cycle" in the United Kingdom is shown in Figure A.2.4. It shows that tyres for ultimate disposal in the United Kingdom arise at four points: manufacturers, casing selectors, retreaders and reclaimers. The situation in the United States, as shown in Figure A.1.1. does not differ markedly, except that the entrepreneurial role of the casing selector does not appear in the United States, and international trade in casings is of no importance there. The United States and Australian tyre industries appear to have similar structures.

Chapter IV

DETERMINANTS OF APPROPRIATE MANAGEMENT POLICIES FOR USED TYRES

The problem of the management of used tyres emerges at the level of retail outlets, regional depots of tyre manufacturers, retreading factories, reclaiming enterprises, and specialist entrepreneurial services. The problem arises either because the used tyres are no longer suitable for retreading, or because there is an excess supply of suitable casings of a particular construction. (This applies, for instance, to cross ply, textile-reinforced tyres in the United Kingdom). The tyres have been concentrated at these depots with little extra social cost, but the actual storage of the tyres represents a visual disamenity and a significant fire hazard. The appropriate disposal route depends upon:

i) transportation and handling costs
ii) costs of disposal process
iii) market for any materials produced by the disposal process.

1. TRANSPORTATION AND HANDLING COSTS

Estimates of transportation and handling costs are particularly difficult to find. Transport costs are to some extent dependent on the form in which tyres are transported. When tyres have been reduced in volume by grinding, splitting or chopping, transport costs per tyre are significantly reduced, but handling and processing costs are increased. Mechanical volume reduction of tyres is particularly energy-intensive, and also requires a significant labour input. The situation in the United States, as in most other countries, varies from metropolitan to rural localities. Representative figures are given in Table 3. From Table 3 and assuming that there are no large reductions in handling costs resulting from transporting ground as opposed to whole tyres, it appears unlikely that the volume reduction achieved by grinding and chopping could be justified in terms of transport cost savings. In practice, most freight companies switch from volume tariffs to weight tariffs if the tyre is first mechanically reduced in volume. This is certainly the Australian experience [6, pp. 25-27]. Based on a trailer with a capacity of 72 m3 and a maximum payload of 18 tonnes, the following payloads could be carried:

- either, a full volume load of 1895 whole tyres, with an estimated total weight of 14.78 tonnes based on an average weight of 205 kg/m3;
- or, a full weight load of 2,308 shredded tyres, based on a volume reduction of 30 % of the original volume, and a shredded density of 585 kg/m3.

Table 3

COLLECTION, HANDLING AND TRANSPORTATION COSTS FOR SCRAP TYRES, USA, 1974 *

(US $)

	Cost per tyre
Collection	0.37
Handling (Loading, unloading)	
manual	0.04
on pallets	0.02
Transportation	
Train 100 miles	0.06
300 miles	0.12
1 000 miles	0.20
Truck 100 miles	0.04
300 miles	0.21
Grinding, chopping	0.10 - 0.25

* Source: 25, p. 15.

On the basis of this volume reduction, the American figures suggest that the extra processing cost would far outweight the savings brought about through volume reduction. Given the Australian estimate of 21 ¢ (Aus.) per tyre for shredding, it seems unlikely that a 35 % volume reduction would be economically justifiable in that country. The situation in Australia has been studied in depth, the conclusions of the study being, inter alia:

i) with the current vehicle sizes and loading regulations, higher drivers wages together with higher capital and operating costs largely offset the advantage of transporting denser cargoes

ii) significant reductions in operating costs could be achieved if vehicles able to carry loads of greater volume were allowed on the roads [6, p. 25, from 22].

In the case of the 18 tonne trailer, holding 22 % more tyres when they are shredded, the conclusion is that there is little, if any, economic advantage in shredding at individual depots prior to long distance haulage. There are also figures available for the costs of tyre collection in Canada. It is estimated that these costs (in an area with a relatively low population density) vary from Can. $13 - Can. $30 per tonne. This represents a cost of 15¢ (Can.) - 30¢(Can.) per passenger tyre. In addition, landfill charges are estimated to be around Can. $5.00 per tonne.

Road transport costs in Australia appear to vary little either with the number of trailer loads or the frequency of the loads. Table 4 shows representative transport costs (including handling costs).

Table 4

ROAD TRANSPORT COSTS FOR SCRAP TYRES, AUSTRALIA *

To Sydney from	Distance (km)	Cost/load (a) (Aus. $)	Cost/tonne (b) (Aus. $)	Cost/tyre (c) (Aus. ¢)	Cost/tyre/km (Aus. ¢)
Newcastle	180	241	16.30	12.5	0.069
Canberra	305	268	18.13	13.9	0.046
Melbourne	920	504	34.10	26.2	0.028
Adelaide	1 435	550	37.21	28.6	0.020

a) Based on 2 hours loading/unloading per trailer load.

b) Based on full 72m3 load of whole tyres weighing 14.78 tonnes.

c) Based on 130 tyres/tonne.

* Source: 6, p. 26.

Rail transport costs would be on a weight basis in Australia. Rail transport is more expensive than road transport, even excluding costs of transporting tyres to and from railway terminals, and the extra handling costs involved. Cost per tyre kilometre decreases more slowly with distance for rail transport. It is concluded that rail transport for whole or shredded tyres is unlikely to be competitive with road transport under normal circumstances. The figures are given in Table 5.

In the Canadian Prairie Provinces, out of a scrappage rate of 30,000 tons of scrap tyres (approximately 3,950,000 tyres) per year, only 12,800 tonnes (1,689,600 tyres) per year are recovered. The main constraint is the transportation costs that would be involved in collecting scrap tyres. The density is reduced by stockpiling by individuals and use by farmers [27]. A further conclusion "that transportation costs can be reduced ... by shredding" suggests that Canadian freight schedules are on a different weight and volume basis than those in Australia. Even so, adding in the costs of shredding, it appears unlikely that the volume reduction achieved will make up for the increased total costs including shredding.

Table 5

RAIL TRANSPORT COSTS, SHREDDED TYRES, AUSTRALIA *

To Sydney from	Distance (a) (km)	Cost/tonne (Aus. $)	Cost/tyre (b) (Aus. cents)	Cost/tyre km. (Aus. cents)
Newcastle	180	16.46	12.7	0.070
Canberra	350	25.06	19.3	0.063
Melbourne	920	47.56	36.6	0.040
Adelaide	1 435	67.66	52.0	0.036

a) Distance by road.

b) Based on 130 tyres/tonne.

* Source : 6, p. 27.

In the United Kingdom, tyre transportation costs by road are estimated to by £0.05 per tonne mile /21, p. 124/. This again is presumed to include handling costs. There is no indication as to whether freight costs for used tyres are less favourable than for new tyres, as it is the case for some scrap materials in some countries. Given the importance of transport costs in determining the feasibility of tyre recycling, this information would be of considerable value.

Costs for transporting tyres to landfill sites have been calculated in some detail in the Federal Republic of Germany /2, p. IV/66/. Transport costs of whole tyres to a landfill site are estimated at 25 DM per tonne, for a trip of 20 kilometres, and 53 DM per tonne for a trip of 50 kilometres. The former trip was for the median distance from the tyre fitters to a collection centre (which may include the collection of several partial loads) and the latter for transporting whole tyres from the collection centre to the landfill site.

Transport costs for tyres in the Netherlands are high. (1) A truck carrying 50 cubic yards of used tyres costs 40 Hfl. per hour, and 1 Hfl. per kilometre. The load would contain 300 passenger car tyres. In other words, total transport costs, assuming tyres on average need to be moved for a one hour journey of 60 kilometres, for the total volume of passenger car tyres generated in the Netherlands would be Hfl. 1,700,000 per annum.

A recent French study /17, pp. 80-81/ has drawn attention to the collection system for used tyres in that country. Whilst emphasising that information on collection and disposal costs for used tyres is limited, it reports collection costs for car tyres at FF104 per tonne (FF0.80 per tyre), and at FF220 per tonne (FF9.90 per tyre) for heavy

1) Material provided in a private communication, Stichting Verwijdering Afvalstoffen, vdK/Rh/882, 3rd August, 1977.

vehicles tyres. The report suggests further that systematic tyre collection could be encouraged by subsidies or preferential freight rates for the transport of tyres.

2. COSTS OF DISPOSAL PROCESSES

Information on the costs of disposal for the various categories of wastes entering the municipal waste stream are notoriously limited. Individual countries estimates of the costs of particular processes are discussed below with reference to used tyres. There are few general evaluations of alternative disposal routes and processes. Amongst the most thorough is that presented by Wilcox [29] in which carefully operated, high quality sanitary landfill "remains the cheapest acceptable waste disposal method", although particular local circumstances (such as the absence of suitable sites in the vicinity) may change this. The suitability of other systems depends on a complex of factors which makes it particularly difficult to establish a ranking. It should be emphasised that local factors tend to weigh heavily in the suitability of particular systems to particular localities. It is also worth noting that capital-intensive waste management technologies do not generally feature prominently in the least-cost rankings.

3. THE MARKET FOR RECLAIMED RUBBER MATERIALS

The economic viability of processes involving the recovery of rubber, or of component materials, from discarded tyres to some extent determines the viability of these processes. The materials that can be recovered include rubber crumb, reclaim ("reconstituted" rubber), and the gas, heavy and light oils, and char that can be obtained on the pyrolysis of used tyres.

In the cases of rubber crumb and reclaim, the market has been weakening over the years due to the falling off in effective demand. The reasons for this are mainly technical: higher, more stringent specifications on automobile tyres in particular have meant that the demand for "secondary" rubber, either as crumb or reclaim, has been falling. Similar, upward revisions in specifications have meant that the demand for secondary rubber in the production of other rubber goods, such as conveyor belts, has also been falling.

The market for the products of tyre pyrolysis has yet to be effectively established in Member countries. The gas is usually retained to keep the process going. The oils are usually sufficiently mixed as to make them suitable only for crude industrial heating applications. The char fraction is rarely of sufficient quality to be used in tyre production, and if it can be so used, it can only be used in the production of tyres for which the specifications are not demanding, e.g. tyres for agricultural and some construction vehicles. These matters are discussed in more detail below.

4. ALTERNATIVE DISPOSAL ROUTES FOR USED TYRES

Figure 2 shows the alternative disposal routes that are technologically feasible either at the moment, or will be in the near future. Each alternative will be examined individually, but prior to this it will be of use to understand the determinants of whether tyres go to retreading or disposal, and, in the latter case, which ultimate disposal route is followed.

The decision as to whether a tyre is suitable for retreading, or whether it should be scrapped, is determined largely by three factors:

i) The state of the carcass. This is of paramount importance. Tyres with damaged side walls, or that have been punctured or slit, cannot be accepted for retreading. Any other physical defects will also automatically cause the tyre to be rejected.

ii) The second determinant of whether a suitable (undamaged) used tyre is retreaded is the type of tyre. This covers both the size of the tyre, the arrangement of the plies, and the type of reinforcement. For example, in the United Kingdom, the demand for retreaded, cross-(bias-) ply textile-reinforced tyres is falling off relatively rapidly. Also, whilst manufacturers often use unusual tyre specifications on new car models, retreaders are rarely prepared to carry a full range of moulds that would enable them to retread every suitable tyre. Hence, tyres sold only in small volumes, or tyres unique to imported cars, are less likely to be accepted for retreading.

iii) The type of vehicle on which the tyre is used also influences the likelihood of a tyre being retreaded.

Automobile tyres are retreaded at most only one (75 % or more not at all). Truck and bus tyres can be retreaded two or three times. Aircraft tyres can be retreaded up to five times. In the United Kingdom in 1974, 36.7 % of all replacement tyre sales for trucks were remoulds. In Germany, the proportion of retreads in the replacement sales market is estimated to be 27 % for cars and 49 % for trucks; in Italy, the figure is 27 % for cars and 50 % for trucks; in the Netherlands, 15 % of replacement car tyre sales are retreads, and 49 % of those for trucks; in Norway and Sweden respectively, 31 % and 15 % of replacement car tyres are retreads, as are 40 % and 30 % of truck tyres. In France, retreaded tyres only represented 8 % of the market for replacement tyres in 1971. However, all heavy vehicle tyres are retreaded at least once, and 80 % are retreaded two or more times [17, p. 61]. In Austria,

35 % of replacement car tyre sales are retreaded (the highest proportion recorded in any Member country [21, pp. 27-28].

There is some evidence on the state of used tyres that have particular wear or damage characteristics. In the Netherlands, a sample of 2 % of the turnover of a used tyre collecting and reclaiming company was taken [5]: 30 % of the sample was cross-ply, radial tyres representing the remaining 70 %. Of the radial-ply tyres, 43 % had been discarded before the tread had been worn to the safe limit. In the case of cross-ply tyres, 50 % had been discarded too early. Thus, for the sample and for the sample period, 45 % of all tyres discarded had been discarded prior to their useful life (as guaged by treadwear) being realised.

Table 6 summarises the reasons for the early discards. The importance of balance and alignment of the wheels, and inflation of the tyre, are clear to see.

Table 6

TYRE DISCARDS PRIOR TO ACHIEVEMENT OF
NORMAL USE BY CATEGORY OF TYRE

(In per cent)

Reason for discard	Cross-ply	Radial-ply
Improper fitting	0.0	0.6
Incorrect wheel balance	15.0	10.2
Cracks and abrasions	2.0	6.2
Poor alignment	19.0	17.0
Incorrect inflation	14.0	9.0

In Denmark, where 30 % of the tyres discarded are retreaded (the proportion varying between automobile tyres and heavy vehicle tyres), similar studies have been made to determine the reasons for tyres being rejected by retreaders [14, p. 14]. 50 % of the discarded tyres were either cross-ply tyres, for which there is no retread market in Denmark, or were "first-generation" radial-ply tyres, which were not retreadable. 15 % of the discarded tyres had defects in the bead wire, or had become greasy on removal from the vehicle. The tyre body of the remaining 5 % had either been physically damaged, or had suffered excessive wear due to being insufficiently inflated.

Tyres which are not suitable for retreading, or which have not been inspected as to their suitability for retreading, must pass into one of the three major tyre scrapping channels: physical applications (including landfill and stockpiling), chemical processes, or energy extraction. The determinants of which of these disposal routes are followed are:

Figure 2 **RECYCLING, RE-USE AND DISPOSAL ALTERNATIVES FOR SCRAP TYRES**

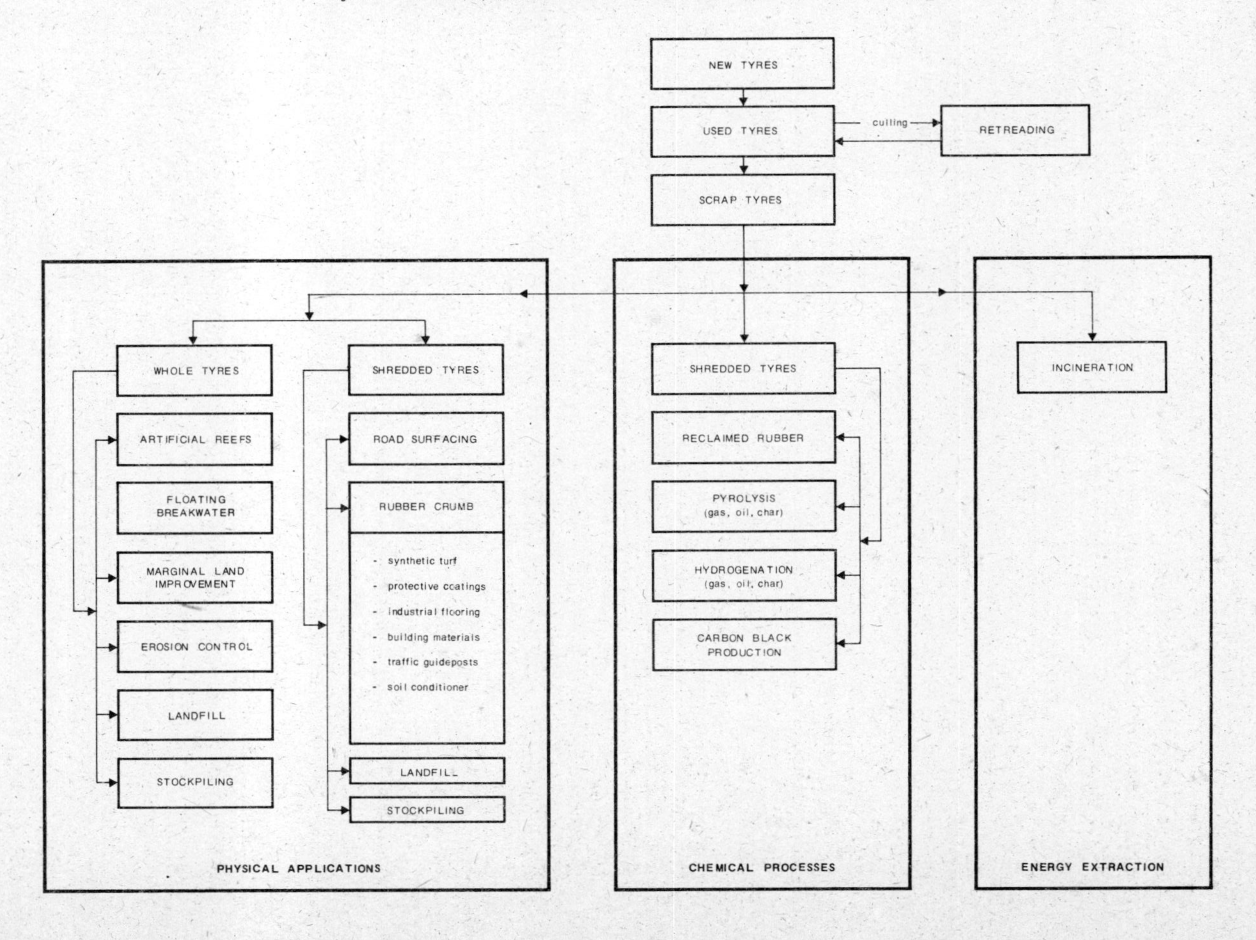

i) the proximity of appropriate facilities

ii) regulations/restrictions on certain forms of disposal

iii) type of reinforcing material (for example, steel-reinforced tyres cannot be accepted for conventional rubber reclaiming processes)

iv) the state of the market for the materials produced by the process, and the flow of social costs and benefits that stems from a particular use for scrapped tyres.

Chapter V

PHYSICAL USES FOR SCRAPPED TYRES

In physical applications, scrapped tyres can either be used whole or shredded. Whole tyres can be directed to landfill, or used for :

- artifical reefs
- floating breakwaters
- marginal land improvement
- erosion control

or they can be stockpiled.

Shredded tyres can also be landfilled or used in road surfacing. As rubber crumb (produced by grinding or by more recent cryogenic means), tyres can be used in the manufacture of:

- synthetic turf
- protective coatings
- industrial flooring
- building materials
- traffic guideposts
- soil conditioners.

1. LANDFILL

In many countries, as is the case for many categories of solid waste, burying the tyres in landfill is the most common method of disposal. In some countries, however, in metropolitan areas, landfill sites are not easily available, and sites further removed from the cities are being sought. With transportation costs rising as the sites used are further away, other disposal options are likely to become relatively more attractive. In Sydney, it is estimated that disposing of tyres in this way costs about 23 Australian cents per tyre, or Australian $30 per tonne for collection and transportation and 2.3 Australian cents per tyre for disposal. The total cost is 25.3 cents per tyre /6, pp. 40-41/. In the USA, collection of tyres, manual handling, transportation over 100 miles by truck, and simple dumping costed between 43.25 cents to 43.50 cents per tyre in 1974. In contrast, collection, manual handling, truck transportation over 100 miles, and disposal into sanitary landfill costed 46 cents to 49 cents per tyre in 1974 [25, p. 15]. The land residual problems of whole tyres in landfill are long-term and persistent. Used tyres are quite resistant to biological attack; they are a fire hazard if dumped too densely; they can leave ground spongy and unfit for building on. Whilst shredding the tyres overcomes many of these problems, it increases the full costs of landfill by approximately one-third in USA and one-half in Australia. It has also been reported that leachate from tyres may contaminate soil and water, although the evidence is slight. Monetary benefits of landfill stem from its being, currently, the least-cost disposal method in most areas (without shredding prior to disposal). Non-monetary costs arise from the fire hazard, health hazard and loss of visual amenity that whole tyres present in many sites.

In the United Kingdom, the problem of disposal is to some extent reduced by disposing of tyres down used mine shafts. It can be calculated from the bottom line of Table 7 that $\frac{96.798}{554.057}$, or 17.5 %, of all tyres discarded in Britain appear to be disposed of by tipping into landfill sites (the majority) or mineshafts (the minority), but only 2.54 % were estimated to go to incineration. This table does not show, however, the increasing shortage of local landfill sites, in parts of England [11, pp. 40-41]. Similar regional problems are arising in other countries. It should be noted that in the evaluation of used tyre disposal by landfill, the land is estimated to have zero opportunity cost. Inasmuch as proper landfill procedure without tyres may enable eventual building development on the site, whereas including tyres may limit the use of the area to recreational development (or some other development that does not include building) this loss of development potential should be allowed for. It is not likely, in any practical sense, however, to alter landfill's position as the least private cost disposal option.

The costs of tyre disposal by landfill have received considerable attention in the Federal Republic of Germany [2, IV/64-71]. Five alternative schemes have been evaluated. All of them involve two transport stages: one from the point of disposal to a collection centre, and the other from the collection centre to the landfill site. Only three of the schemes will be mentioned here, however. The first involves controlled landfilling, with tyres being shredded at the site. The costs of this scheme are 129DM per tonne, or 0.90DM per passenger car tyre. In the second scheme, involving shredding at the collection centre and the transport of shredded tyres to the landfill site (thus using second stage transport more efficently), the costs are 203DM per tonne, or 1.35DM per passenger car tyre. A third scheme, involving the provision of special landfill sites for tyres only would cost, it is estimated, 140DM per tonne, or 1.00DM per passenger car tyre. As a standard of comparison, the costs of disposing of whole tyres in a mixed waste landfill site are estimated at 13.50DM per tonne. Certain external costs are not included, nor are the costs of transporting the tyres in the mixed wastes to the landfill site. The costs of shredding are estimated to be 39DM per tonne (approximately 0.27DM per passenger car tyre).

In Denmark, the charge for accepting tyres into landfill was reported to be 46 D.Kr. per tonne [14, p. 9]. The rate is 150 % of that for household waste.

The French report comes down strongly in favour of using controlled landfill as a management alternative for used, non-retreadable tyres [17, pp. 75-76]. The report, in fact, stresses that, in that country, it is the only waste tyre management alternative through which the volume of waste tyres generated can be handled, at least in the short-run. Under average conditions, and allowing for excavation to four metres, approximately 25,000 tonnes of shredded tyres could be disposed of in this way per hectare of ground.

Whilst there is no direct evidence for the costs of landfilling tyres in Japan, there is considerable indirect evidence to suggest that the costs are so high as to make other used tyre management alternatives much more attractive than in other Member countries. Of the tyres that were collected in that country in 1974, 90 % were recycled, either as retreaded tyres or as reclaimed material, and only 10 % were _either_ incinerated _or_ landfilled [18, p. 7]. This contrasts strongly with other Member countries, where the proportion of used tyres going to landfill is extremely large.

The greatest drawback of the landfill disposal option is that the resources included in the tyre are thrown away. It has been estimated, for example, that about 32 litres of crude oil are required, on average, to make a tyre. Only 10 % - 15 % have been lost when the tyre becomes unroadworthy. The opportunity cost of the resources thrown away in this way, without the rubber crumb or the chemical constituents or the heat being reclaimed, and without society being able to realise some of the non-monetary benefits of tyres being used for artificial reefs (for example), might be a significant cost that is not accounted for in least-cost disposal estimates. A full evaluation of the landfill disposal option for tyres would take into account the following costs and benefits. From this, it can be seen that the least-cost aspect of landfill concentrates entirely on costs borne through factor markets. When other, non-monetary factors are taken into account, intuition suggests that the attractiveness of landfill for used tyres may be reduced.

Table 7 *

ESTIMATED DISPOSITION OF USED TYRES IN THE UNITED KINGDOM, 1971-1972

(Tons)

From \ To	Manufacturer	Wholesale depot	Retreaders/ remoulders	Casing selector	Reclaimers	Commercial enterprise	Export	Incineration	Tipping	Farmers	Bon-fires	Total
Tyre manufacturers	446	59	1 027	3 792	630	532	0	5 110	7 981	0	0	19 577
Wholesale depots	6 666	131	6 437	648	1 546	10	0	83	890	72	3	16 940
Retail outlets	8 528	46 184	115 235	26 231	1 487	6 349	31	1 971	5 008	332	3	224 349
Retreaders/remoulders	4 639	3 813	4 135	18 132	18 961	1 000	1 437	6 223	27 085	160	0	85 585
Casing selectors	14 193	115	34 124	5 518	22 697	4 020	15 877	295	53 483	2 160	0	152 584
Reclaimer	0	0	0	0	0	0	0	0	25	0	0	25
Wholesaler/retailer	13 485	124	25 673	3 836	3 521	184	19	387	2 326	18 261	1	67 720
Import	0	0	116	612	0	0	0	0	0	0	0	728
Total	47 957	50 426	186 747	58 767	48 842	12 095	17 364	14 069	96 798	20 985	7	567 508 / 554 067

This table should be read as follows: reading across a particular line shows the disposition of used tyre arisings from a particular source, the source identified at the beginning of the line. For example, line 3 shows that retail outlets have a total estimated arising of used tyres of 224 349 tons. Of these, 8 528 tons went (presumably as backloading) to the manufacturer, 46 814 tons went to wholesale depots, and so forth. Thus, the column totals represents acceptances by particular outlets. Row totals represent arising. The difference between arisings (567 508) and acceptances (554 057) is partly due to survey error, but partly due also to miscellaneous disposal routes, including rogue disposal on a small scale.

* Source: 11, p. 41.

Costs		Benefits	
Monetary:	1. Handling 2. Transport 3. Collection 4. Tipping	Monetary:	None
Non-monetary:	1. Reduced development potential of the landfill site 2. Value of "thrown away" resources embodied in tyre	Non-monetary:	1. Reduced fire hazard 2. Reduced health hazard 3. Reduced visual disamenity (relative to storage of whole tyres)

2. FLOATING BREAKWATER AND ARTIFICAL REEFS

Both of these alternatives have been used to dispose of tyres. In the United States, the Bureau of Sport, Fisheries and Wildlife (BSFW) has been experimenting with artificial reefs made from used tyres since 1965. Scrap materials are used because of their low cost. Junked cars and ship hulks, as well as scrap tyres, have been used. Rubber tyres have been found most satisfactory on the four criteria used to evaluate reefs:

- life expectancy of the material
- surface area
- encrustration characteristics
- variability of reef design.

BSFW estimates that artificial reefs could absorb all scrap tyres from the USA for several decades. (Transport considerations would preclude this, however). The estimated reef construction cost is US$2.24 per tyre (1) which, together with the collection, handling and transport costs to the reef site of 45 cents per tyre would raise the total cost to US$2.69. The other consideration is the non-monetary benefit of increased recreational fishing facilities. If reefs were to be used to stimulate commercial fisheries, then the benefits would increase significantly.

Table 8 shows the estimates of reef-building costs in the United States, using different designs of tyre unit. Costs of tyre acquisition

1) Made up of:

Ballasting and connection material	$0.22
Labour	$0.57
Transportation to site	$1.45

Tyres were presumed to be delivered free to the dockside.

Table 8

COST PER MULTI-TYRE UNIT FOR REEF BUILDING, USA *

(US $ per tyre)

Description of tyre unit	Material (d)	Labour	Transport	Total
12 tyre, triangular array, on reinforcing rods and weighed with concrete (a)	0.49	0.75	2.90 (e, f)	4.14
8 tyre set, chained (a)	0.64		1.00 (e, g)	1.64
8 tyre stack, stainless steel or plastic bands (a)	0.17	0.20	0.14 (e)	0.51
4 tyre stack, stainless steel or plastic bands (a)	0.07	0.25	0.10 (e)	0.42
1 tyre, 7 kg concrete between sidewalls (a)	0.07	0.19	0.08 (e)	0.34
8 tyre stack, speared on reinforcing rod and weighed with concrete (a)	0.16	0.20	0.14 (e)	0.50
4 tyre stack, speared on reinforcing rod and weighed with concrete (b)	0.11	0.89	0.56 (e)	1.56
3 tyre, side by side in concrete base (c)	0.68	2.05	1.35 (e, h)	4.08

a) Costs based on study by the US EPA, Department of Fisheries, and the National Tyre Dealers and Retreaders Association.

b) Costs based on private reef project.

c) Costs based on State supported reef project.

d) Based on free delivery of tyres to a dockside staging area.

e) Based on transport by barge to reef at a cost of $700 per day for a tow vessel.

f) Based on a part load of test units.

g) Includes labour.

h) Includes loading costs.

* Source: 6, Table 8.1., p. 36.

are not included, however. Tyre reefs have an advantage over other tyre disposal options in that construction can be started or stopped with little adverse effect.

Floating breakwaters are constructed by partially filling tyres with foam rubber, and lashing them together in modular bundles. They have excellent energy absorbing characteristics. Cost estimates vary: a breakwater 180 metres long, with a life expectancy of 20 years, would require 600 man-hours of labour and materials costing $30,000. The same breakwater with a 5-year life-expectancy would require the same labour but the materials cost would fall to $12,000 [6, from 3]. In Australia, the system is thought to be economically feasible but unlikely to absorb significant quantities of used tyres.

3. EROSION CONTROL AND MARGINAL LAND IMPROVEMENT

There is some evidence of scrap tyres having been used for erosion control. They have been used lashed together to form large, flexible covers over banks of dams and rivers to assist stabilise vegetation. For stabilising sand-dunes, a mat, 25 feet wide, would need 87,000 tyres per mile of shoreline [9, p. 45]. Such a use could absorb a limited number of tyres for a limited period of time. One disadvantage of whole tyre use in this form is the visual disamenity caused by large volumes of tyres spread over the surface of the land.

The burial of tyres to raise the level of swampy land is a further use for whole tyres. Tyres which are laced together are stacked into freshly-dug trenches in the swamp, and then the earth is replaced on top [6, p. 39]. This compresses the tyres and raises the ground above high water level. The land is reported to consolidate quickly and can soon be used for recreation. Such a scheme in Sydney is reported to cost 24.5 Australian cents per tyre for tyres used in this way, this being the cost of collection, transportation and disposal; costs of digging and infilling are not given [6, p. 39]. This scheme now uses approximately 200,000 of the scrap tyres arising in Sydney, but is thought only to be able to do so for another few years. The environmental impact of this use of whole scrap tyres seems minimal. There may be some danger of the tyres working their way to the surface in the long-run. Costs of using tyres in this way in the USA are estimated to be between 10-15 US cents per tyre, in addition to the collection, handling and transport costs already reported, of about 45 US cents per tyre.

As tyres are not significantly subject to biological decomposition when used intact for land reclamation, erosion control, reef building, and so forth, there is little evidence of any adverse environmental effects, apart from the visual disamenity that some of these uses may cause. The real concern for this general category of tyre use is that potentially valuable materials are irrevocably thrown away. This argument is only sound, however, if the expected costs of recovering the materials are less than the expected market value of the reclaimed materials together with disposal cost savings. Otherwise, these whole tyre uses may be socially the most beneficial.

4. SHREDDED TYRES

Tyres are shredded for several uses, including pre-shredding prior to incineration or prior to addition to landfill. Tyre shredding requires a great deal of energy, due to the construction of the tyre. Shredding costs have been estimated at A.$0.20 per tyre in Australia, between US$0.10 and US$0.25 in USA, and at 0.27DM in West Germany. Whilst increasing the cost of the process, shredding can, in fact, reduce some of the problems associated with alternative means of disposal, for example, problems associated with incomplete combustion of whole tyres, or with the flexing of tyres in landfill. Shredding also puts tyre stockpiling on a more viable footing than when whole tyres are involved. In the USA, it has been estimated that the entire annual output of the nation could be stored, to a height of 50 feet, in a space of only 150 acres per year. Alternatively, ten regional sites of 300 acres each could handle the accumulated (shredded) output of scrap tyres from the USA for 20 years [9, p. 83]. The advantage of stockpiling scrap tyres in this way is that, when the petroleum products from which synthetic tyre rubber is made move into a situation of excess demand [also in about 20 years], the stock of shredded tyres would be a bank from which synthetic rubber constituents could be reclaimed (presuming that the technology of doing this has improved by then). The disadantage is that storage of scrap rubber provides as great a visual disamenity as the storage of most other forms of scrap. In addition, unless proper preventative arrangements were taken, the fire hazard that such a storage depot represent would be great. In Australia, a total area of 130,000 m3 of storage space has been estimated to be necessary. Presuming a 10-year storage period, and 20 strategic sites around the country, this would require that each site be 0.065 hectares, with the shredded tyres stored to a height of 10 metres [6, p. 46]. The inherent flexibility of a tyre stockpiling scheme commends it greatly. With the careful choice of sites with low opportunity costs (few alternative uses) and minimal loss of visual amenity, improvements in recycling technology and increases in the prices of raw materials (particularly petroleum feedstock) should make the policy socially beneficial as well as economically profitable. No figures are to hand on the cost of stockpiling used tyres, although it is believed that one major stockpiling scheme is already in existence, in Saskatchewan, Canada.

The Federal Republic of Germany stockpiles tyres, although this appears to be a market-related activity as there is no clear government programme to encourage tyre storage. Of the (approximately) 340,000 tonnes of tyres discarded in that country each year, approximately 100,000 tonnes (or about 30 % of total discards) go into storage [2, p.IV/27]. This storage appears to be short-run, and a significant proportion of the tyres stored in this way may simply be an inventory awaiting disposal. No other Member country appears to be pursuing active policies at the moment to build a stockpile of tyres, awaiting technical developments in pyrolysis or chemical reclaim processes.

5. RUBBER CRUMB

Rubber crumb (crumbed vulcanised rubber) can be produced mechanically or by cryogenic crushing. Uses for rubber crumb include:

- various uses in road-building and repair
- incorporation in synthetic turf
- use in protective coating
- use in building materials
- use as a soil conditioner.

Many of these applications are still in the development stage, and there are no reliable cost data on them. One exception to this is a Canadian study, reporting that comminution of scrap tyres by cryogenic crushing to produce rubber crumb (for incorporation into asphalt for road surfacing) costs 6.4 Canadian cents per pound from a central facility situated in Regina. If a series of provincial facilities were established for the Canadian Prairie Provinces, the costs would be 6.6 Canadian cents per pound in Edmonton, 7.3 Canadian cents per pound in Regina, and 16 Canadian cents per pound in Winnipeg [27, p. 9]. The higher costs for the regional facilities are probably due to economies of size and scale in the currently-available cryogenic processes. Other sources, however, question the economic viability of cryogenic crushing "unless there is a cheap source of low temperature liquid gas" [21, p. 41].

There is some evidence that gives comparative costs for cryogenic crushing. Requiring 0.9kg. of liquid nitrogen to produce 1 kg of scrap rubber crumb of a specified grade (50 mesh), apparatus is available in the following sizes:

Capacity (kg liquid N per hour)	Equipment Cost (£)
15 - 150	2 100
25 - 600	2 500
75 - 1 000	4 200
100 - 3 000	7 000

Equipment which could handle 5,000 tyres per eight-hour day had a capital cost of £78,000 /21, pp. 54-56/. No information is available, however, on operating costs for this process. Information from Australia sets the cost of cryogenic comminution at A$1.00 per tyre /6, p. 79/ against A$0.20 per tyre for shredding. The advantage of the cryogenic process is that it enables steel-reinforced tyres to be handled for crumb manufacture. It is not at all obvious, however, that the additional cost involved is justified in terms of the extra solid waste management problem that steel-reinforced tyres represent.

The Swedish government has awarded a grant to build a tyre recovery plant using the cryogenic separation technique. The plant is being built under an energy-saving ordinance; the 50 % grant amounts to 8,550,000 Swedish krone. 75 % of the product of the plant is expected to be of a grain size less than 0.4 mm. The plant is being built to handle 8,000 tonnes of tyres per year. (1)

In Federal Republic of Germany, 90,000 tonnes of the tyres discarded go to shredding. This represents about 26 % of the total volume of tyres discarded /2, p. IV/27/.

In Denmark, as in other countries, the viability of reducing used tyres to rubber crumb depends on the price received for the crumb. Very fine crumb commands a premium. Table 9 shows the prices received for varying grades of crumb in Denmark, and the method by which the crumb is produced.

Table 9 *

RUBBER CRUMB PRICES AND METHOD OF PRODUCTION, DENMARK

Method	Grade	Price (D. Kr./tonne)
Mechanical	5-10 mm	60
Mechanical	1.6 mm	260
Mechanical	0.6 mm	700
Method Unknown	< 10 mm	240
Cryogenic comminution (no cleaning)	0-30 mm	790

* Source: 14, p. 16.

1) Personal communication, Swedish Environmental Protection Agency.

The use of rubber crumb in road construction and repair

There have been experiments with the use of rubber crumb in road construction in many countries over approximately 40 years. The best documented experience since 1967 has been that which originated in Arizona, USA, as a result of collaboration between the Arizona Department of Highways and a private company (Sahuaro Petroleum and Asphalt Company, Phoenix, Arizona). The procedure involves the addition of 25 % to 30 % of granulated treat rubber between 16 and 25 mesh reclaimed from discarded automobile tyres to a hot asphalt mixture at 300°F-450°F. The resultant mixture is a tough and elastic binder with less susceptibility to temperature changes than conventional asphalt /15/.

Amongst the benefits claimed for this process are the following when the mixture is applied as a seal coat:

i) the process prevents reflection cracking from the substrate pavement because of its flexibility and the interlaced particles of rubber discourage the propagation of craks. (The Arizona Department of Highways and the Company both emphasise, however, that the seal coat construction is not a solution to all types of bituminous concrete failure).

ii) The asphalt-rubber cement is waterproof, and hence the moisture in the road subgrade is stabilised and diminishes the tendency for local failures in the surface.

iii) The "temperature susceptibility" of the binder is reduced, reducing the tendency of the binder to bleed in hot weather or crak due to shrinking or flexing in cold weather.

iv) Used as a method of road repair, on cracked road surfaces, it does not raise the "profile height" of the pavement and does not, therefore, necessitate raising the height of the curbs.

v) Asphalt-rubber seal coat pavements are predicted to have a life of 10 years, (1) compared to a maximum of 5 years for conventional asphalt chip seal coat.

In addition to these advantages in the maintenance of roads, savings in landfilling (or other disposal), costs should be attributed to the scheme. Other savings that should be attributed to this method of road maintenance are the materials and energy savings. The reason for this is that, in comparison with a one-inch "standard asphaltic overlay", the rubberised asphalt process uses only 1/3 of the amount of asphalt, 1/4 of the amount of burner fuel used (to heat and pretreated the chips) and 1/3 of the amount of fuel used to lay the asphalt, compared to conventional asphalt cement processes. (2) If the

---

1) Inspection by the author of strips that had been in use in Phoenix, Arizona, for between 5-10 years confirmed that these surfaces were in good condition. See also 26, pp. 49-54.

2) Figures provided by Sahuaro Petroleum on the basis of independent tests run by the Arizona Department of Transportation.

claims of some local authorities that have used rubberised asphalt prove to be correct - that a one inch application of rubberised asphalt is effective in reducing reflective cracking and fatigue to the same extent as a three to four inch overlay of conventional asphalt; the savings in materials and energy would be even greater [15, p. 24].

The cost of using asphalt-rubber as a seal coat was reported to be between 50 US cents and 70 US cents per square yard in 1973, this figure including traffic control, cover stone, pre-heating and pre-coating of cover-stone, asphalt-rubber binder, kerosene additive, placing, and finally, rolling. Applying an extremely crude inflation adjustment to this figure, the current cost would be between 70 and 98 cents per square yard. Against this cost figure would have to be set savings on maintenance costs due to improved durability of roads. By contrast, in Australia, it has been estimated that addition of rubber crumb to asphalt mixes in road construction would reduce asphalt use by 0.2 % to 3.0 %. Again, added to materials savings costs would need to be the reduction in maintenance costs that would follow from the reduced frequency of maintenance. There is no data on the use of rubber crumb in sports surfaces, protective coatings, etc. Table 10 indicates that rubber crumb is potentially a substitude in many applications, but the evidence is that so far the "potential" market for crumb in these uses simply has not developed. The reason lies obviously in the cost of rubber crumb.

Table 10

APPROXIMATE COSTS OF RUBBER CRUMB AND MATERIALS FOR WHICH IT MAY SUBSTITUTE *

Material	Application	Approximate cost (A.$/cu.m.)
Rubber crumb	Road surfacing	200
Asphalt	Road surfacing	100
Vermiculite	Spoil conditioner	100
Polyurethate	Upholstery padding	65
Polystyrene	Insulation, protective padding, etc.	25
Mineral wool	Insulation, protective padding, etc.	20
Fibreglass	Insulation, protective padding, etc.	20

* Source: 6, p. 48.

The use of rubber as an additive to asphalt for road surfaces has been studied extensively in Canada [27, p. 278]. In that country, consumption of natural rubber goods in 1974 was 291,000 tonnes. If this can be taken as an approximation of the volume of rubber waste generated per year, it represents less than half of one per cent of the municipal waste stream. The replacement of natural rubber by synthetic elastomers means that the proportion of natural rubber in the waste stream is falling.

Tyres represent 70 % of the elastomers ("synthetic rubbers") discarded in Canada. Whilst a significant reclaim and re-use industry already exists in Canada, it reprocesses only a small fraction of the scrap generated. The study concludes that the waste tyre management effort in the Canadian Mid-West ought to have two thrusts:

i) an attempt to reduce tyre scrappage rates by encouraging the manufacture and use of longer-life tyres (discussed below)

ii) an attempt to utilise the scrap tyres discarded by using them, in the form of cryogenically-produced rubber crumb, as an additive to asphalt used in road construction.

It is this second route for used tyres that we shall concentrate on for the time being.

The Canadian study (henceforth known as the "Wardrop Study") outlines a cost-benefit analysis of the use of rubberised asphalt application to roads in the Prairie Provinces. The exercise identifies both costs and benefits in present value terms. The benefits identified are the savings in materials costs that will be achieved by replacing current types of road treatment by rubberised asphalt, plus any net savings achieved by reclaiming materials rather than discarding them irretrievably.

The cost of the rubber crumb is included separately from the other costs of the seal coat. This may enable a sensitivity analysis to be undertaken to determine the effects on the scheme of different price regimes for rubber crumb. However, the costs of the rubber processing facility are not included in the analysis. Instead, a procedure is adopted whereby "the cost of rubber has been allocated as the minimum price at which a central processing facility would be expected to sell the material". As long as this minimum price includes the full costs of the capital needed for producing the crumb, and is not simply a variable-cost covering, short-run price, then the alternative procedure is acceptable. If, however, the price only covers variable costs then it will be an entirely inadequate measure, and the procedure will produce a serious distortion in the analysis (significantly underestimating costs). The early net benefits of the scheme would be particularly exaggerated.

The cash flow statement for one of the proposed plants is presented in table 11, for information. The cost figures show that there is a strong incentive for the plant to be run at or as near to maximum capacity as possible, as there are significant reductions in average total cost to be gained thereby. The figure for the minimum price of rubber crumb (Can. $0.30 per kilogramme) covers these costs for all but the first two years of operation. Similar figures are attributed to other plants with different capacities. There is no indication of the current price of rubber crumb in the Prairie Provinces.

The benefit-cost analysis for the use of rubberised asphalt is given in Table 12. The basis for the cost figures needs to be detailed:

i) the reduction in disposal costs attributed to benefits does not include an allowance for the reduction in visual disamenity that this disposal option will achieve (nor is this allowed for anywhere else);

ii) full employment is assumed throughout the study - no allowance for employment-generating effects is made. Similar considerations apply to reductions in seasonal employment resulting from the reduced maintenance on roads;

iii) the impact on the petroleum refining sector attributable to the reduced use of asphalt is not counted as a cost to the project;

iv) the possibility that petroleum-based products are underpriced has not been allowed for; market prices rather than shadow prices are used throughout. More generally, no attempt is made to correct any of the distortions resulting from the real world diverging from the perfectly competitive model of economic theory;

v) discount rates of 5, 10 and 15 per cent were used to test the robustness of the model to fluctuations in the social opportunity cost of capital;

vi) all net benefits are in real terms (not taking inflation into account);

vii) the investigation is made for a period of 10 years commencing in 1979.

The net benefits indicated by the study are Can.$159,000, Can.$ 118,000 or Can.$89,000, depending upon whether the discount rate is set at 5 %, 10 % or 15 %. The scheme appears to be entirely viable, regardless of the real opportunity cost of capital that is allowed. There is, however, a certain problem in using this study as a criterion for decision-making; the costs and benefits of other waste tyre disposal schemes are not allowed for in the study. Most are dismissed as being ineligible. It would, however, be a more thorough guide to tyre management policy if all the options had been similarly evaluated.

The possibility of using rubber crumb in road construction has also been studied in other Member countries, without its being met with unanimous approval. In Switzerland, (1) an estimate made in 1975 concluded that, under optimal conditions (i.e. supposing that rubber crumb is added to bitumen at a rate of 10 % W/W for the replacement of

1) Personal communication, Swiss delegate to the Waste Management Policy Group.

Table 11 *

CASH FLOW STATEMENT - CENTRAL PLANT LOCATED IN REGINA

(Can. $1,000)

	1	2	3	4	5	6	7	8	9	10	11
Expenditures											
Plant equipment	550	0	0	0	0	0	0	0	0	0	0
Installation	220	0	0	0	0	0	0	0	0	0	0
Building and ancillary equipment	1,375	0	0	0	0	0	0	0	0	0	0
Miscellaneous fixed costs	0	40	40	40	40	40	40	40	40	40	40
Labour	0	55	58	65	72	79	80	86	93	100	106
Liquid nitrogen	0	90	148	278	407	537	561	667	800	934	1,040
Power	0	1	2	4	6	7	8	9	11	13	14
Collection	0	48	79	148	218	287	300	356	428	499	556
Total direct costs	0	235	327	535	742	950	989	1,158	1,372	1,586	1,755
Interest on working capital	0	8	9	13	16	19	20	23	26	30	33
Interest during construction	161	0	0	0	0	0	0	0	0	0	0
Interest on capital	0	231	251	264	258	231	180	121	39	0	0
Principal repayment	0	0	0	60	273	507	596	824	385	0	0
Total	2,306	473	587	872	1,290	1,707	1,785	2,126	1,822	1,616	1,788
Working capital	0	98	136	223	310	396	412	483	572	661	732
Debt outstanding	2,306	2,511	2,644	2,584	2,311	1,805	1,209	385	0	0	0
Revenue											
Sales											
Saskatchewan	0	153	337	524	709	895	742	850	959	1,068	1,177
Manitoba	0	0	0	25	50	76	101	126	176	227	252
Alberta	0	0	0	205	410	616	821	1,026	1,295	1,564	1,770
Toronto	0	114	114	114	114	114	114	114	114	114	114
Scrap steel	0	1	2	4	6	8	8	9	11	13	15
Total sales	0	268	453	872	1,290	1,707	1,785	2,126	2,556	2,986	3,327
Long term debt	2,306	205	134	0	0	0	0	0	0	0	0
Total	2,306	473	587	872	1,290	1,707	1,785	2,126	2,556	2,986	3,327
Cash flow	0	0	0	0	0	0	0	0	734	1,371	1,539
Rubber produced (1000 kgs)											
Sold to Saskatchewan	0	510	1,125	1,746	2,364	2,982	2,472	2,835	3,198	3,561	3,924
Sold to Manitoba	0	0	0	84	168	252	336	420	588	756	840
Sold to Alberta	0	0	0	684	1,368	2,052	2,736	3,420	4,318	5,214	5,899
Sold to Toronto	0	455	455	455	455	455	455	455	455	455	455
Total rubber produced	0	965	1,580	2,969	4,355	5,741	5,999	7,130	8,559	9,986	11,118
Cost of rubber produced ($/kg)	0.00	.49	.37	.27	.23	.21	.20	.18	.17	.16	.16
Plant capacity 600 tyres/hr.											
Liquid nitrogen 1.1kgs./kg. of rubber $0.85/kg											
Collection transportation, and shredding $.050/kg.											
Interest on capital 10 %											
Sales of rubber $.30/kg. fob Regina or Toronto											
Sales of steel scrap at $.02/kg.											

* Source: 27, p. 150.

Table 12 *

BENEFIT-COST ANALYSIS FOR THE USE OF RUBBERIZED ASPHALT IN THE PRAIRIE PROVINCES

(Can.$ 1,000)

Year a)	2	3	4	5	6	7	8	9	10	11	Total
<u>Costs</u>											
<u>Saskatchewan</u>											
Seal Coat (exc. rubber)	332	734	1,136	1,538	1,940	1,608	1,844	2,080	2,316	2,552	
Equipment	100	200	100	100	100				50		
Rubber ($.30/kg)	153	338	524	709	895	742	850	959	1,068	1,177	
Total	585	1,272	1,760	2,347	2,935	2,350	2,694	3,039	3,434	3,729	24,145
Rubber (1000 kg)	510	1,125	1,746	2,364	2,982	2,472	2,835	3,198	3,561	3,924	
<u>Alberta</u>											
Seal coat (exc. rubber)			515	1,030	1,545	2,060	2,575	3,298	4,031	4,536	
Equipment			200	150	100	150	150	100	100	100	
Rubber ($.30/kg)			205	410	616	821	1,026	1,295	1,564	1,770	
Total			920	1,590	2,261	3,031	3,751	4,693	5,695	6,406	28,347
Rubber (1000 kg)			684	1,368	2,052	2,736	3,420	4,316	5,213	5,900	
<u>Manitoba</u>											
Seal coat (exc. rubber)			55	110	165	220	275	385	495	550	
Equipment			50			50				50	
Rubber ($.30/kg)			25	50	75	100	125	175	225	250	
Total			130	160	240	370	400	560	720	850	3,430
Rubber (1000 kg)			84	168	252	336	420	588	756	840	
<u>Prairie Provinces</u>											
Seal cost (exc. rubber)	332	734	1,706	2,678	3,650	3,888	4,694	5,763	6,842	7,638	
Equipment	100	200	350	250	200	200	150	100	150	150	
Rubber ($.30/kg)	153	338	754	1,169	1,586	1,663	2,001	2,429	2,857	3,197	
Total	585	1,272	2,810	4,097	5,436	5,751	6,845	8,292	9,849	10,985	55,922
Rubber (1000 kg)	510	1,125	2,514	3,900	5,286	5,544	6,675	8,102	9,530	10,664	
<u>Benefits</u>											
<u>Saskatchewan</u>											
Saving in oil treatment	600	1,200	1,800	2,400	3,000	1,200	1,200	1,200	1,200	1,200	
Saving in asphalt conc.	2,057	4,550	7,043	9,536	12,029	12,029	12,029	12,029	12,029	12,029	
Total	2,657	5,750	8,843	11,936	15,029	13,229	13,229	13,229	13,229	13,229	110,360
<u>Alberta</u>											
Saving in asphalt conc.			4,647	9,267	13,941	18,588	23,235	24,082	24,929	24,929	143,618
<u>Manitoba</u>											
Saving in asphalt conc.			678	1,356	2,034	2,712	3,390	4,116	4,842	4,890	24,018
Total <u>Prairie Provinces</u>	2,657	5,750	14,168	22,559	31,004	34,529	39,854	41,427	43,000	43,048	277,996
Present disposal costs ($.01/kg)	5	11	25	39	53	55	67	81	95	107	
Total benefits	2,662	5,761	14,193	22,598	31,057	34,584	39,921	41,508	43,095	43,155	278,534
Benefits less costs	2,077	4,489	11,383	18,501	25,621	28,833	33,076	33,216	33,246	32,170	222,612
<u>Present value</u>											
Discounted at 5 %	1,977	4,072	9,835	15,226	20,087	21,509	23,517	22,487	21,444	19,752	159,906
Discounted at 10 %	1,888	3,708	8,549	12,636	15,911	16,262	16,968	15,512	14,096	12,804	118,334
Discounted at 15 %	1,807	3,394	7,490	10,583	12,734	12,456	12,437	10,862	9,442	7,946	89,151

a) Assumes facility start-up in January 1979.

* <u>Source</u>: 27, p. 100.

the seal coat of all main roads), only 2,750 tonnes of crumb per year would be required. This corresponds to about 3,500 tonnes of used tyres, out of the 8,500 to 15,500 tonnes disposed of by landfill every year. Even though the total length of roads to which the rubber asphalt could be applied has increased since 1975, and even though this use of scrap tyres would allow absorption of at least 20 % of the tyres that would otherwise be wasted, it seems that certain technical considerations (uncertainty in the choice of the optimal rate of crumb to take account of differences between summer and winter conditions, possible difficulties when trying to completely recycle the seal coat containing rubber crumb) have so far militated against introduction of this practice in Switzerland.

Chapter VI

CHEMICAL USES FOR USED TYRES

Chemical uses of scrap tyres can be thought of as uses in which the controlled chemical treatment of scrap permits the recovery of certain original or related chemical constituents. All processes involve the initial reduction of the whole tyre to smaller sized pieces.

1. RECLAIMING (1)

In most Member Countries, reclaim production has gradually reduced since World War II, as virgin polymer became more easily available and prices fell (in both nominal and real terms). Some statistics on rubber production and use in Member countries are shown in Annex 4. The information available does not reflect the impacts of higher raw materials (mainly petroleum feedstock) and energy prices upon reclaim production since 1974. There is some evidence, however, that the net effect of these changes has been to increase the demand for reclaimed rubber in some Member countries.

The market for reclaim depends upon its cost of production and upon its qualities relative to virgin rubbers and plastics, for which it is (to some extent) substitutable. The relatively small proportion of reclaim used in new tyre production is due to a technological problem. With existing blending technology, reclaim cannot be used in proportions greater than 1-2 % for high performance tyres. The problem will increase if governments make tyre performance standards more stringent. Thus, technical difficulties in substitution may prevent widespread adoption of reclaim in tyre production, even with the virgin material/reclaim price ratio continuing to favour reclaim. In the market for conveyor belts, stricted standards have similarly reduced the demand for reclaim. Both in the United Kingdom and in Australia, the market for reclaim is so declining that the industry is contracting. In Australia, an additional reason for the industry's decline is reported to be the import of reclaim at prices lower than those of

1) Useful technical details on the various processes for reclaiming rubber from used tyres are contained in 19.

indigenous producers. At a level of 600 tonnes in 1975, this represented approximately 25 % of Australia's total use of reclaim.

Whilst costs of production or reclaim are not directly available, the proportionate allocation of costs between various operations is shown in the figures in Table 13, whereby comminution is effected either mechanically or cryogenically. (1)

The production of rubber reclaim involves, however, considerable air and water pollution, and yields considerable quantities of solid waste, in the form of metal bead wires (and increasingly metal reinforcing materials) and fabric. The process produces highly contaminated waste water containing oil, grease, sulphur derivatives, softeners and suspended solids. The effluent has a high chemical oxygen demand, and often has a pH value significantly different from neutrality. Depending upon the process used, significant quantities of metal sludge containing metal chlorides, hydrolised fibre and fine rubber particles, are also generated.

In the Federal Republic of Germany, it is estimated that reclaim production can absorb up to 10 % of the waste tyres generated in that country. For other natural and synthetic rubber products, the proportion is estimated to be 25 %. However, in Germany, as in most other countries, the price of reclaim has often been in excess of the price for virgin natural rubber. In 1974, the natural rubber price was around 1.60 DM/kilogram, whereas, over the preceeding three year period, the price of reclaim had fluctuated between 1.15 DM/kg and 3.30 DM/kg. It is this problem of the price of reclaim often exceeding that of natural rubber which militates strongly against its more widespread adoption.

In the Netherlands, (2) reclaim is produced mainly from natural rubber, and, as such, the process is only able to absorb an increasingly smaller proportion of the waste tyre flow. Moves are being taken, however, to further the manufacture of reclaim from tyres made mainly of synthetic rubber.

Denmark imported 456 tonnes of reclaim in 1975, and manufactured 15,000 tonnes at home [14, p. 26]. In Denmark, the relative prices of reclaim, new natural rubber and new synthetic rubber were very much in favour of reclaim in 1975 (see Table 14). This price advantage appears to have resulted in a higher proportion of reclaim being used in rubber products manufactured in Denmark.

In Canada, there is now only one rubber reclaim manufacturing plant, and reclaim is used only to 4.71 % of the elastomers used in high quality, high performance tyres, but may be used up to 25 % in lower quality tyres. There is little evidence of reclaiming increasing its contribution to waste tyre management in Canada.

---

1) Personal communication with R.G. Norman, United Reclaim Company Limited, Liverpool 14, England.

2) Material provided in a private communication, Stichting Verwijdering Afvalstoffen, vdK/Rh/882, 3rd August, 1977.

Table 13

COSTS OF PRODUCING RECLAIM RUBBER USING EITHER MECHANICAL COMMINUTION OR CRYOGENICS AND MECHANICAL COMMINUTION

1) Operating costs (expressed as a percentage of total cost) for the production of rubber reclaim from scrap car tyres by mechanical comminution and the reclaimator process.

Cost item / Operation	Cracking	Separating	Grinding	Reclaiming	Refining	Packing	Totals
Materials	2.61	-	-	19.99	-	1.21	23.80
Electricity	1.97	3.81	2.38	3.81	5.92	-	17.88
Waste disposal	0.39	2.21	-	-	-	-	2.61
Labour	5.82	5.79	4.25	4.15	6.56	3.88	30.49
Maintenance	1.84	2.09	2.00	3.06	1.96	0.49	11.43
Depreciation	0.69	0.72	0.40	0.50	0.51	0.51	3.32
Other costs	2.31	2.30	1.20	1.42	1.53	1.72	10.47
Totals	15.63	16.92	10.23	32.93	16.48	7.81	100.00

2) Operating costs (expressed as a percentage of total cost) for the production of rubber reclaim from scrap car tyres by cryogenic comminution and the reclaimator process.

Cost item / Operation	Cryogenic	Grinding	Reclaiming	Refining	Packing	Totals
Materials	13.79	-	19.72	-	1.19	34.70
Electricity	0.90	1.12	3.76	5.84	-	11.62
Waste disposal	1.68	-	-	-	-	1.68
Labour	11.13	3.61	4.09	6.47	3.83	29.13
Maintenance	0.97	0.77	3.02	1.93	0.48	7.17
Depreciation	0.19	0.35	0.49	0.50	0.50	2.03
Other costs	8.09	0.97	1.40	1.51	1.70	13.67
Totals	36.75	6.82	32.48	16.25	7.70	100.00

Table 14 *

PRICES FOR NEW RUBBER AND RECLAIM, DENMARK, 1975

(D.Kr./kg.)

Type	Price
Reclaim	1.75
Synthetic rubber	4.10
Natural rubber	4.00 - 6.50
Retreading mixture before revulcanising	6.50

* Source: 14, p. 28.

The French report also notes that the market for reclaim is shrinking in that country. Unlike other countries, however, the French seem to think that the market for reclaim could be expanded once more (although the report does not offer a secure basis for this forecast) [17, pp. 72-73].

2. DESTRUCTIVE DISTILLATION, CARBON BLACK RECOVERY AND HYDROGENISATION (1)

It is possible, using various chemical processes, to recover the chemical constituents of scrap tyres. Two of the processes (destructive distillation, and carbon black recovery, or carbonisation) are forms of pyrolysis. Hydrogenisation is a process of chemical synthesis. All processes are largely in the developmental stage, and cost estimates are preliminary. Hydrogenisation involves the addition of hydrogen to rubber to make chemicals from which new elastomers can be produced.

In pyrolysis, the ultimate constituents of tyres (carbon, hydrogen, and ash, together with small quantities of nitrogen and sulphur) are yielded in chemically complex oils and gases, and a solid residue. Depending on the operating temperature, the proportion of oil, gas and residue can be varied. High temperature (900° C) pyrolysis yields large quantities of residue, much of which is carbon black (which in turn represents 1/3 to 1/4 of the total constituents of a tyre by weight). Lower temperature pyrolysis yields large quantities of oils, mostly olefins, aromatics and naphthenes. However, the commercial value of these processes has yet to be determined. Many of the gases and oils produced by destructive distillation can be used on site for their fuel value, but the commercial viability is still in doubt. As the IR & T report states :

1) Useful and comprehensive technical information on the various pilot pyrolysis plants that have been constructed is available in 19, pp. 120-175.

"The proportion of gases, oils and residues that would be produced would be determined by economic considerations. The survival or abandonment of a particular method of constituent recovery would depend on capital and operating costs. Within the limitations of the methods which are currently feasible, the decision to manufacture a particular product would depend on the demand for that product at the price for which it can be produced profitably. Because developmental work on these methods is still in an early stage, it is very difficult to determine accurately the probably installed costs for systems using them. None of these methods can yet operate at a profit." [9, p. 30].

The costs of waste tyre pyrolysis

The costs of installing and operating a new technology are difficult to estimate at the developmental stage. The costs presented below are estimates, in almost every instance, from pilot plants, and will most probably be subject to considerable revision under commercial operating conditions.

The Commission of the European Communities has attempted to estimate the capital and operating costs of pyrolysis. (1) As a first approximation, capital costs are estimated to be between US$500,000 and US$1,000,000 per ton per hour capacity. The simplest, and least capital cost, technology is that using large pieces of tyres and operating at relatively low temperatures (300°C). The total operating costs of the processes surveyed fell between US$65 and US$90 per ton of waste tyres treated. The commercial yields from pyrolysis are expected to be:

i) oil, valued at circa US$80 per ton as a replacement industrial fuel oil (suggestions that it could be used as rubber industry feedstock have been discounted)

ii) carbonaceous residue, or char, whose value is estimated to be US$30 per ton.

It should be emphasised again that these calculations on the value of pyrolysis products are not based on commercial experience.

The Carbon Development Corporation of Michigan, USA, have correctly identified carbon black as potentially the most profitable yield from pyrolysis [6]. The process is described as follows:

... the shredded tyres ... are "pyrolyzed in a rotary hearth furnace, maximizing carbon yield by charring the tyres in a non-oxidising condition to 900°C. The gases evolved are burned to produce superheated steam at approximately 480°C. The steam is

---

1) 17, pp. 126-127. The methodology for estimating the costs is as follows: operating costs have been estimated on the basis of US$6.00 per hour for labour, US$0.04 per Kwh, and US$120 per tonne for fuel. These figures "reasonable for France in 1977" were estimated by the French consultant to the CEC study.

used to pulverize the char in a fluid energy mill creating the particle size and surface chemistry necessary for carbon black. This pulverized char is then coated, to protect the active surface from oxygen, and then pelleted for final use". [6, p.1].

For a metropolitan area of 1 million people in the United States, and presuming current use and disposal patterns, a plant of 2,000,000 tyres per year capacity would be needed to process all the scrap tyres that such an area would produce each year. (1) The yield from such a plant is expected to be approximately 11,600 tons (26,000,000 pounds) of carbon black. The capital investment required for such a plant would be US$4,000,000 (15.4 US cents per annual pound capacity carbon black). Operating costs plus depreciation are estimated to be 9.5 US cents per pound carbon black. The price expected for the carbon black produced is 13 UScents per pound, or an expected profit of 3.5 US cents per pound of carbon black. (Again, the assumption that carbon black produced could be sold at the same price as virgin carbon black is untested, and questionable).

Information on a pyrolysis plant that has been operating in Hamburg provides probably the most comprehensive evaluation of a pyrolysis project currently available [10]. The cost and revenue estimates are shown in Table 15. A few points should be noted from this table:

i) 25 % of the gross revenue is represented by the acceptance fee, which, if pyrolysis becomes widely used, is likely to fall (and may well become a negative element if the demand for used tyres increases sufficiently).

ii) The price of carbon black is presumed to be that received for the lower grade product used for agricultural equipment tyres (although none has been sold for this purpose yet).

iii) The value for gas is an imputed value, as the gas generated during the process is used to keep it going.

iv) More generally - although the size of plant is small, the figures presented by Janning et al. show little evidence of lower costs for higher capacity plants.

v) The advantage of this plant is that, besides waste tyres, the system can accomodate plastic wastes, waste oils, and waste paper. (2)

vi) The tyre pyrolysis products from the plant have been further refined, the oil fraction undergoing extensive separation. It is thought that the sale of the benzine fraction (and some of the other fractions) separately may increase the economic viability of the process. (2)

---

1) The presumption that all the waste tyres generated in such an area would go through only one facility is unrealistic.

2) Information obtained during a visit to the plant in January, 1979.

Table 15 *

COSTS AND REVENUES FROM THE PYROLYSIS OF SCRAP TYRES
(For a 15 tonnes per day, 4,000 tonnes per year plant drawing tyres from a 30 km. radius)

Costs (per annum) (1)	DM
Running and repair costs	75,000
Administration and personnel	240,000
Collection	140,000
Interest foregone on investment (10 % on DM 1,500,000)	150,000
Depreciation (20 % on DM 1,500,000)	300,000
Annual costs	905,000
Revenue (per day)	
Acceptance fee	1,285
Gas	405
Benzine (aromatics)	1,020
Fuel oil	378
Carbon black	1,575
Metal	60
Daily gross revenue	4,723
Annual equivalent revenue	1,317,200
Annual costs	905,000
Annual net revenue	412,000
Net revenue per tonne	98

*) Source: 10.

1) All figures are rounded.

Pyrolysis of waste tyres is a rapidly developing technology. There still do not appear to be plants operating on a commercial scale, although the chances are that there will be within one or two years. The problem of obtaining a high quality char is still a major one. The rising cost of petroleum feedstocks for producing elastomers is, together with the increasing concern about throwing away a potentially valuable resource, the main incentive for improving the process. A successful tyre pyrolysis industry, together with a policy of scrap tyre stockpiling, could be one way to preserve the tyre industry in the future from severe shortages of feedstocks. However, a great deal of further research is required on these techniques. Cost-sharing arrangements between government and industry, as in Birmingham, England, or between government, universities and industry, as in Hamburg, Federal Republic of Germany, appear to be favourable ways of undertaking this research providing that the results are made fully available.

Chapter VII

ENERGY EXTRACTION

Conventional tyre incineration has several obvious drawbacks:

- large quantities of acrid black smoke are produced
- high levels of sulphur dioxide are also produced
- the high calorific value of the tyre is wasted.

Modern developments in tyre incineration have been towards:

- ensuring complete combustion of the tyre
- producing refractory materials able to withstand the high temperatures of tyre incineration (2000° C)
- ensuring adequate control over particulate and sulphur dioxide emission produced by tyre incineration.

The French report [17, p. 63] identifies the main pollutants resulting from burning tyres in the open air. The proportions are summarised in Table 16.

Table 16

MATERIAL COMPOSITION OF A PASSENGER CAR TYRE AND POLLUTANTS PROCUDED BY COMBUSTION IN THE OPEN AIR *

Composition	Pollutants
45 % polymers	oxydes of nitrogen
27 % carbon black)	carbon monoxide
1 % stearic acid)	and carbon dioxide
11 % oils	carbon dust
1,4 % zinc oxide	zinc oxide
1 % sulphur, steel and textiles	oxides of sulphur

* Source: 17, Table 4, p. 63.

Most developments have also been towards ensuring that the heat generated is not dissipated, usually by using incinerators that function in conjunction with a waste heat boiler. Steam generated can be used:

- to aid on site in the vulcanising process for new tyre rubber
- in central heating
- for electricity generation (after superheating the steam).

Tyre incinerators are unlikely to be used for general power generation, as insufficient tyres are produced in any one location to feed such a facility. Energy from tyres is more likely to be used for heating public buildings, or for industrial purposes. Costs vary with the capacity of the furnace, the efficiency of incineration and the efficacy of its emission control. Because of the combustion properties of rubber, tyre furnaces are likely to operate only at 30 % efficiency (whereas coal, for which tyres are to some extent a substitute, can be burned at around 60 % efficiency). There are estimates from USA that the cost of generating power from tyres would be US$0.49 per million B.T.U. (exclusive of collection costs but including amortisation costs of capital equipment and operating costs of storage, handling, preparation and emission control) compared to US$0.35 per million B.T.U. for coal. Nevertheless, excluding coal, tyres are cheaper than other fossil fuel energy sources [9, pp. 26-28].

In the Federal Republic of Germany, two per cent of the waste tyres are incinerated [2, p. IV/43]. In Denmark, the costs of incineration and disposal are estimated to be 120 D.Kr. per tonne (as compared to 40 D.Kr. per tonne for mixed municipal waste) [14, p. 8]. In the Netherlands, the incinerators in Rotterdam charge municipalities or private tyre distributors 30-35 H.Fl. per tonne for incineration in addition to which the costs of transport to the incinerators have to be borne by the clients. (1)

In 1974, costs of simple incineration of tyres in USA were estimated to be US$0.20 to US$0.40 per tyre. In Australia, costs were estimated to be A.$1.00 per tyre for collection and disposal. An Australian study of the costs of tyre disposal with steam [6, p. 74] estimated costs of power generation with used tyres to be A.$13 per tyre.

> "The study report concludes that the annual value of steam produced (A.$75,000) would be approximately equal to the annual operating and maintenance costs (A.$79,000). On the basis of a capital cost of A.$780,000, interest at 10 % per annum, and an amortisation period of 10 years, the annual cost would be A.$127,000. If the plant is to operate so as to recover all costs, at 400,000 tyres per year, a subsidy of 33 Australian cents per tyre would be needed". [6, p. 74].

---

1) Information provided by SVA: 29/10/1977.

Chapter VIII

TYRE LIFE EXTENSION

The alternative to management of scrap arisings in various ways is to reduce the volume of waste arisings at source. For tyres, the problem of source reduction can be handled by encouraging more careful maintenance of tyres, by expanding tyre retreading, and by producing tyres whose initial useful life has been significantly increased.

1. IMPROVED MAINTENANCE

As mentioned above, standards of tyre maintenance are uniformly low in Member countries. Underinflation, fast acceleration, fast cornering and severe braking all cause tyres to wear out before their useful life, under proper maintenance and more careful driving conditions, has been reached. As well as reducing the life of the tyre, this also tends to reduce the number of tyres going to retreading. Approaches that might encourage more thorough maintenance include:

i) imposing a deposit on tyres, the refund being conditional on the tyre being accepted for retreading (this would increase the proportion of passenger car tyres retreaded)

ii) more clear advertisements, funded by governments, on appropriate inflation pressures, and the personal costs of severe driving and improper maintenance

iii) part exchange agreements between retreaders and tyre users, similar to those run by automobile electrical parts dealers in some countries.

2. INCREASED RETREADING

To some extent, the capacity to increase the use of retreaded tyres rests in the private car sector. In several countries, business have agreements with tyre companies or with retreaders to take all suitable tyres in part exchange for new ones. This is particularly true in the United States. It has been estimated that tyre trade-in when the tyre still had 1/16" tread would increase the number of tyres in retreadable condition by over 35 % in the United States. There is also a standard in that country that requires a tread wear indicator to show when only 1/16" tread remains.

The potential for increased retreading is significant. Provided that retreads can be produced at a competitive price relative to new tyres, and that their standard can be maintained, the following measures may encourage greater retreading [6, p. 19]:

i) more strict policing of worn tyres
ii) ensuring that retreaded tyres are produced to rigid standards
iii) increased use by governmental organisations of retreaded tyres (procurement requirements/legislation)
iv) reducing the sale of cheap, "non-retreadable" tyres (i.e. pricing policies of tyre companies)
v) limiting variations in tyre specifications on new car models.

3. IMPEDIMENTS TO INCREASED RETREADING

There are several factors which work against increasing the proportion of retreaded tyres in total tyre sales in Member countries. Amongst these factors are:

- technical difficulties in retreading certain tyre types;
- the high market share accruing to lower quality tyres;
- the cost of retreading relative to the cost of manufacturing a new tyre;
- the pricing and sales policies of new tyre companies;
- concerns about the safety of retreaded tyres.

Each of these impediments is discussed below, to a greater of lesser extent.

There are two sets of problems that may be regarded as technical problems associated with retreading. The first relates to the tolerance between the mold and actual tyre size for radial tyres. For conventional tyres, this tolerance is 0.5-0.88 inches. For radial tyres, the tolerance is 0.25 inches or less. Thus, more molds are required to retread radial tyres than conventional, bias-ply tyres. With the increase in the proportion of radial-ply tyres in the tyre stock, and the fact that most retreading enterprises are only small concerns, the increase in capital costs that retreading radial tyres involves has probably contributed to a reduction in the rate of retreading, at least in the United States [26, p. 57].

The other technical impediment to retreading also relates to "first generation" radial tyres. The problem is known as "belt-edge lift", and concerns the tendency for the metal plies to raise at the edges once the used tyre has had any excess tread rubber removed. This prevents the new tread from adhering properly to the tyre body.

For more recent radial tyres, this problem has, apparently, been overcome. (1)

The existence of lower quality tyres on the market is also a disincentive to retreading. They generally lack the "structural strength" required for retreading, and their use "dilutes the stream of potentially re-usable carcases" [6, p. 16].

The cost of retreading varies significantly. An average figure for the United States was US$6.45 per tyre in 1973, with the retreaded tyre selling at $12 to $14. There are no figures to hand for other countries.

The advantages of increasing the proportions of retreaded tyres, and the gains thereby in reducing the volume of waste tyres going into the solid waste stream, are often offset by the pricing policies of new tyre dealers. Most Member countries have reported price wars between new tyre companies reducing the selling price of the cheapest tyres below the unit cost of retreading tyres to a satisfactory standard. In the Netherlands, for example, in late 1977, the costs of retreading a car tyre were around 12 guelders. Cheap, new tyres could be purchased for less. This does not always hold, however. In Denmark, in 1976, the prices of new and retreaded tyres were as shown in Table 17. (The figures were not gained by a statistically sound procedure: they are presented for illustration).

Table 17 *

PRICES OF NEW AND RETREADED TYRES, DENMARK, AUGUST 1976

Model	New steel radial (krone)	Retreaded tyre (krone)	Savings assuming that retread lasts for	
			75 % of new tyre life	100 % of new tyre life
Morris 850	155	103	11 %	34 %
Volvo 144	292	179	18 %	39 %
International truck	1 839	707	49 %	67 %

* Source: 14, p. 21.

In Germany, 27 % of the total volume of tyre discards is collected for retreading, but more than 20 % of these are found unsuitable and returned to landfill or incineration. The market price for carcases suitable for retreading in 1974, varied between 2 - 5 DM per car tyre to 100 DM per truck tyre [2, p. IV/24].

1) Information provided to the author by Mr. P. Taft, Secretary, The National Tyre Dealers and Retreaders Association, Inc., Washington, D.C., USA.

The other aspect of increasing the volume of retreaded tyres in the replacement market is the concern that this may lead to a higher rate of tyre failure resulting in social costs of accidents that could outweigh the waste management gains that the policy was designed to achieve. On the other hand, both BIPAVER (the European Association of Tyre Retreaders) and many Member Countries are eager to emphasise that quality control in the retreading industry has been considerably improved recently.

The United Kingdom Department of Transport has provided updated figures on a survey on the costs and benefits of replacing retreads by original tread tyres. (1) The Department emphasises strongly that both the sample size and the non-randomness of the data limit the policy relevance of the study. They have notwithstanding, kindly allowed the information to be included in this report. The detail is included in Annex 5. The conclusions of this work are that:

> "... unless there is evidence of a significant increase in the relative accident risk of retreads it seems reasonable to expect that the costs of replacing retreads by original tread tyres would outweight the safety benefits to be gained by a factor of at least 3."

If anything, evidence from Europe indicates that the relative accident risk or retreaded tyres is falling, and that the gains from replacing original tread tyres by retreads outweighs the safety costs by a factor of at least 3.

4. INITIAL TYRE LIFE EXTENSION

Over the years, the maximum life of tyres has increased from 10,000 miles to 20,000 miles, to 40,000 miles, and now, reportedly, to 80-100,000 miles. There has been, however, some industry resistance to developing these tyres on a commercial basis. As there has been a history of industries generally retarding the adoption of product-life-lengthening innovations, the case for considerably longer-lived tyres bears greater examination.

A study in the United States /25/ (2), has investigated thoroughly the feasibility of manufacturing a 100,000 mile (160,000 kilometres) tyre and marketing it. The technical feasibility of such a tyre is beyond question: existing truck tyres achieve 115,000 miles (185,000 kilometres) in current use, before the first retreading. A 100,000 mile

1) The original report is from the United Kingdom Department of the Environment, ESD 174/32, 18th November 1977. The update Retreads vs. Original Tyres, RSG/EH C16/19, 23rd August 1978.

2) This section of the report draws heavily on this study and on conversations with Dr. Westerman and the project officer, Dr. Haynes Goddard, of USEPA in Cincinnati, Ohio.

Chapter IX

WASTE TYRE MANAGEMENT: EVALUATION OF THE OPTIONS, AND RECOMMENDATIONS

Over the past few years, many Member countries have displayed increasing concern over the management of discarded tyres. In the United States, concern has shown itself both at Federal, State and City level. The Federal government has considered, at a conceptual level, the application of product charges to tyres. The relevance of the Resource Conservation and Recovery Act 1976 (PL94-580) to waste tyre management, particularly the procurement requirements relating to the content of reclaimed rubber in Federally procured products, has received a good deal of attention [see 30, for example]. At the State level, both California and Kansas have considered the introduction of a charge on tyres, although the California proposal was defeated. Also, the Department of Environmental Protection in the State of Connecticut has drafted plans for waste tyre management including a scheme to stockpile waste tyres [1]. At the City, or metropolitan area level, the Metropolitan Service District of Portland, Oregon, has developed, since 1973, a comprehensive scrap tyre disposal programme [20]. The programme, designed to prevent the illicit dumping of scrap tyres in the metropolitan area of Portland, is reproduced in Annex 6.

In the United Kingdom, local authorities are required to charge a "reasonable fee" for the acceptance and disposal of all industrial waste, including tyres from tyre dealers and retraders. Sweden and Norway are considering extending their deposit/refund scheme on car hulks to tyres. France is considering imposing a product charge on tyres [17, pp. 83-86]. In the Federal Republic of Germany, special sites have been designated for the acceptance of scrap tyres for disposal.

The study has examined all the options currently available to manage the scrap tyres generated each year, and to reduce their volume. The conclusions and recommendations that stem from it all have a common philosophy: that waste tyres be regarded more as a resource to be recovered, as well as a form of waste to be efficiently disposed of or reduced at source.

1. REDUCTION OF THE VOLUME OF SCRAP TYRES

A number of options are directed primarily at the reduction of the volume of tyre wastes to be disposed of: retreading, use of high quality tyres, introduction of longer-life tyre.

Current obstacles to the more widespread adoption of retreaded tyres for replacement purposes on private cars should be identified, and the feasibility of removing these be examined. This would involve measures such as quality control and inspection; advertising the benefits of properly retreaded tyres more fully; examining carefully tyre companies pricing policies with respect to low quality, low mileage tyres; and aiding research to encourage the retreading of steel-belted radial tyres. Also research on the safety aspects of retreaded tyres should be encouraged.

The reasons for the lower market penetration of high quality tyres in the replacement tyre markets for private vehicles ought also to be identified, and the feasibility of removing these be assessed. Amongst the obstacles to more widespread adoption may be: consumer resistance (as is the case with long-life exhausts that can be ordered, at marginal extra cost, for most new automobile purchases, but which are rarely taken up); problems in financing the purchase of more expensive types; ignorance of the higher expense of a policy of purchasing short life, low quality tyres over the life of the car.

Although high quality retreads and high quality original tyres are competitors in the replacement tyre market, substituting them for low quality tyres would achieve significant reductions in tyre waste in the long run.

Moreover, informal conversations with Member countries and with representatives of tyre organisations suggest that the 100,000 mile or 160,000 kilometre tyre is technically feasible, and that prototypes have been produced and tested. However, there is no evidence that such a tyre is likely to be marketed. It is therefore recommended that countries assess, for their own case, the contribution that the introduction of a 100,000 mile tyre on the market would make to waste tyre management and, should the results prove favourable, that countries develop mechanisms by which tyre manufacturers would be prompted to market the longer-life tyre and automobile owners be encouraged to take it up.

2. RESOURCE RECOVERY AND DISPOSAL OPTIONS

Tyres will, however, eventually be scrapped, and they will have to be handled and disposed of. The increasing shortage of landfill sites militates against recommending this as a future policy measure. The costs of proper landfill, with tyre shredding, are high, and the alternative of using special sites for tyres only is also expensive. The practice in some countries of disposing of whole tyres down disused mineshafts is likely to wane as the more readily accessible sites are used up. Also, the uncontrolled disposal of tyres in these shafts presents a significant risk of fire, and consequent air pollution albeit in sparsely populated areas.

There are currently viable means of encouraging the recovery of rubber from scrap tyres, and it is recommended that the feasibility of using rubber crumb in road surfaces, sports surfaces and similar applications be more thoroughly examined.

Policies of storing shredded tyres in safe installation with no adverse environmental impacts should also be evaluated in relation to possible future technologies. The technology of pyrolysis does not yet yield products that can compete, either in terms of quality or of price, and experience in pyrolysis of general domestic wastes does not suggest that developments will be more favourable in the near future. However, safe storage as a means of waste management is used for chemical or hazardous wastes, with the future option of reprocessing or disposal. A similar policy option for waste tyres appears worthy of further investigation.

Research into pyrolysis ought to be continued and encouraged, this research being directed towards the recovery of products with a value, and of a quality, tested in the industrial market place.

3. POLICY ALTERNATIVES

Measures to promote management options aimed at a reduction in scrap tyre arisings, either by source reduction or through resource recovery, comprise a mixture of economic incentives, regulatory requirements, and public sector participation. Among economic incentives to be considered are:

i) a product charge on tyres, partly refundable if the tyre is accepted for retreading (the yield of the charge possibly being used to aid in municipal waste tyre management schemes);

ii) a disposal levy to be charged on each tyre sold, either at the point of purchase or disposal;

iii) removing excise taxes on retreaded tyres, and imposing graduated taxes on new tyres, the level of the tax being inversely proportional to durability and retreading characteristics of new tyres [25, p. 25].

These various measures are aimed to increase the volume of tyres retreaded, and reduce the volume of low-mileage tyres scrapped. Regulatory measures could include:

i) more rigorous tyre inspection
ii) more severe prosecutions for unsafe tyres
iii) licensing of tyre disposal depots
iv) speed restrictions on inter-city highways
v) encouragement for the production of longer-lived tyres, and discouragement for the sale of short-lived tyres which are not likely to be suitable for retreading.

Government involvement could include:

i) education programmes on proper tyre maintenance and advantages of using better quality tyres
ii) funding of research into recovery of chemical constituents from used tyres by chemical means
iii) the use of retreaded tyres and long-life tyres on government vehicles.

The options are numerous, and these are only examples. The feasibility of imposing a product charge on tyres should be examined thoroughly, however. The purpose of such an instrument is clear: the tax, being spread over a longer mileage, would be proportionally lower for longer-life tyres, which present a lower annual pollution load. Similarly retreaded tyres, which have been withdrawn from the disposal stream, should bear a lower charge altogether or no charge at all. The effect of the instrument would be to encourage the substitution of high quality retreads or original longer-life tyres for low quality tyres.

In summary, one fact stands out clearer than any others on the basis of the information available: the ultimate aim of used tyres management policy should be to achieve, as soon as possible, the large reduction in the volume of scrap tyres to be disposed of that can be obtained by implementing the tyre life extending options: increased retreading; the more widespread adoption of currently-available 60,000 kilometre tyres (which, with careful used and maintenance, have been known to last for up to 120,000 kilometres); and the introduction of a longer-life, 160,000 kilometre tyre. In the short-run, to allow for the variations in costs and benefits of the various disposal and recovery options, flexibility should be the rule for any measure which is to be adopted.

4. RECOMMENDATIONS

Having considered the findings of the present report, and the policy options for the management of used tyres, the Waste Management Policy Group of the OECD Environment Committee has adopted the following recommendations:

- that OECD Member countries adopt appropriate measures to promote a more widespread use of currently available options for, and to support research and development into, extending tyre life; two main options are to be considered for this purpose:
  - encouraging an increased use of retreaded tyres (supported by adequate quality control in the retreading industry);
  - encouraging the production, marketing and use of long life tyres;
- that OECD Member countries encourage research into materials and energy recovery from scrap tyres, in association with an examination of the feasibility of storing tyres pending the implementation of new economic recovery techniques; and that they encourage the development of practical uses for recovered materials such as, among others, rubber use in road construction and repair;
- that OECD Member countries consider the use of economic instruments as particularly appropriate means of implementing a comprehensive policy for the management of used tyres.

YEAR HOLDINGS
1967-
VOLUME HOLDINGS

Annex 1

MARKET SHARES FOR DIFFERENT TYPES OF TYRE IN SELECTED MEMBER COUNTRIES

The rate at which the radial ply tyre penetrated the American market is reflected in Tables A.1.1 and A.1.2.

Table A.1.1 *

TYRE SALES BY TYPE IN THE U.S.A., 1968/1969

(percent of sales)

Year	Radial	Bias-ply	Belted bias
1968	1.6	92.4	6.0
1969	2.1	64.9	31.0
1970	2.7	32.3	65.0
1971	3.4	26.6	70.0

* Source: 9, p. 11a.

Table A.1.2 *

PASSENGER CAR TYRE MARKET SHARE BY TYPE OF CONSTRUCTION, U.S.A., 1972-1990

(per cent)

Year	Radial		Bias-ply		Belted bias (1)	
	Original equipment	Replacement	Original equipment	Replacement	Original equipment	Replacement
1972	6	7	16	52	78	41
1975	31	21	10	46	59	33
1980	35	25	5	30	60	45
1985	38	30	2	15	60	55
1990	40	30	0	10	60	60

1) Belted-bias ply tyres are primarily fibreglass reinforced.

* Source: 12, p. 174.

Table A.1.3 shows that the radial ply tyre has taken a larger share of the market in Canada than in the United States. It may be, however, that this is due to a higher percentage of total tyre sales being original equipment on European and Japanese cars that are imported into Canada.

Table A.1.3 *

PASSENGER CAR TYRE MARKET SHARE BY TYRE CONSTRUCTION, CANADA, 1976

(per cent)

Type	Share of market for		Share of total market
	Original equipment	Replacement	
Radial	60	29	36
Belted bias	20	33	36
Bias ply	20	38	28

* Source: 23, p. 83.

Table A.1.4 indicates the situation for Great Britain. It can be seen that the radial ply tyre (an increasing percentage of which are steel reinforced) is taking an increasing proportion of the market, and a significantly greater proportion than in North America.

Table A.1.4 *

PASSENGER CAR TYRE MARKET SHARE BY TYRE CONSTRUCTION, UNITED KINGDOM, 1970-1976

(per cent)

Year	Radial	Cross-ply
1970	37	63
1971	45	55
1972	51	49
1973	62	38
1974	68	32
1975	74	26
1976	80	20
1980	94	6

* Source: 21, p. 13.

In Sweden, the radial ply car tyre represented more than 50 % of the replacement sales by domestic producers, and approximately 90 % of the replacement sales of imported car tyres. For lorries, the situation is slightly different: the radial tyres have not penetrated the domestic tyre market to the same extent (about 30 %), or the market for imported replacement lorry tyres (about 77 %). The situation is clearly shown in Table A.1.5.

Table A.1.5 *

REPLACEMENT TYRES SOLD DIVIDED INTO DIFFERENT TYPES, SWEDEN, 1974

(000's)

	Car tyres			Lorry tyres		
	Diagonal	Textile-radial	Steel-radial	Diagonal	Textile-radial	Steel-radial
4 Swedish producers	753	450	444	126	12	40
15 importers	188	302	805	40	10	127
Total	941	752	1 249	166	22	167

* Source: Personal communication, Technical Department, Municipal Division, Statens Naturvardsverk, Sweden, February, 1978.

Table A.1.6 shows total sales of replacement tyres in Sweden for the same year.

Table A.1.6 *

REPLACEMENT TYRES SOLD IN SWEDEN, 1974
(This investigation covers 90 % of the market)

	Car tyres number x 1,000
4 Swedish producers	1,647
15 importers	1,295
Total	2,942
130 rubber reclaimers (estimated figure)	800
Grand total	3,742

* Source: Personal communication, Technical Department, Municipal Division, Statens Naturvardsverk, Sweden, February, 1978.

In the Federal Republic of Germany, production of tyres was 520,000 tonnes per annum in 1975 of which 65 % were car tyres, 25 % for trucks and 10 % for other vehicles. The market penetration of the steel-belted radial tyre in Germany is shown in Table A.1.7 for the years 1973 and 1980.

Table A.1.7 *

MARKET SHARE OF STEEL-BELTED RADIAL TYRES IN THE FEDERAL REPUBLIC OF GERMANY, 1973, AND FORECAST SHARE, 1980

Vehicle type	Steel-belted radials, 1973 (%)	Steel-belted radials, 1980 (%)	Percentage change
Cars	68	95	+ 40
Small trucks	73	96	+ 32
Large trucks and trailers	58	84	+ 46
Total			+ 39

* Source: 2.

Annex 2

THE TYRE CYCLE IN MEMBER COUNTRIES

The I.R. & T. American passenger car tyre flow model identifies seven sectors in the American tyre market /9, pp. 175-180/:

i) tyre manufacturer
ii) automobile manufacturer
iii) tyre retail
iv) automobile retail
v) tyre retreader
vi) consumer
vii) ultimate disposal outlet.

The use of this model to estimate the volume of waste tyres generated in the United States is shown in Annex 3.

The tyre cycle in Holland is shown in Figures A.2.1 and A.2.2. The importance of distinguishing between the routes for various categories of tyres is clearly reflected by these figures. Instruments or policies designed to reduce a particular flow of waste for car tyres, for example to reduce the volume going to landfill, would not be expected to achieve the same target reduction for truck and bus tyres. The German aggregate tyre cycle is shown in Figure A.2.3. Whilst the tyre cycle in the Federal Republic of Germany is broadly similar to that of other countries, particular details are worth noting. The main feature is the high proportion of tyres in Germany which are disposed of via shredding and controlled landfill techniques (44 % of the total volume ultimately). This proportion appears to be higher than for most other Member countries. Figure A.2.4 shows the tyre cycle in the United Kingdom.

Figure A.2.1* THE TYRE CYCLE FOR TRUCK, 'BUS, AND OTHER HEAVY EQUIPMENT TYRES, NETHERLANDS, 1975

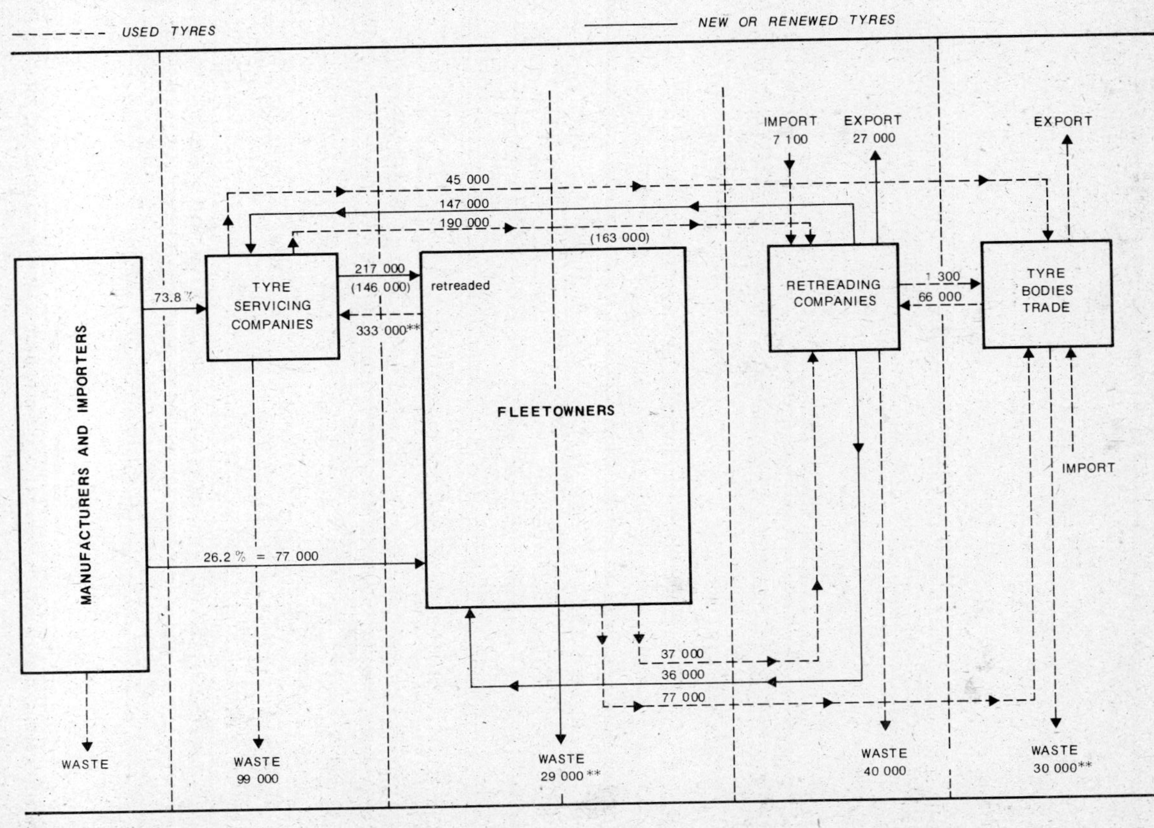

* Source : 24, p. 11.

** Calculated.

Figure A.2.2* **THE AUTOMOBILE TYRE CYCLE, NETHERLANDS, 1975**

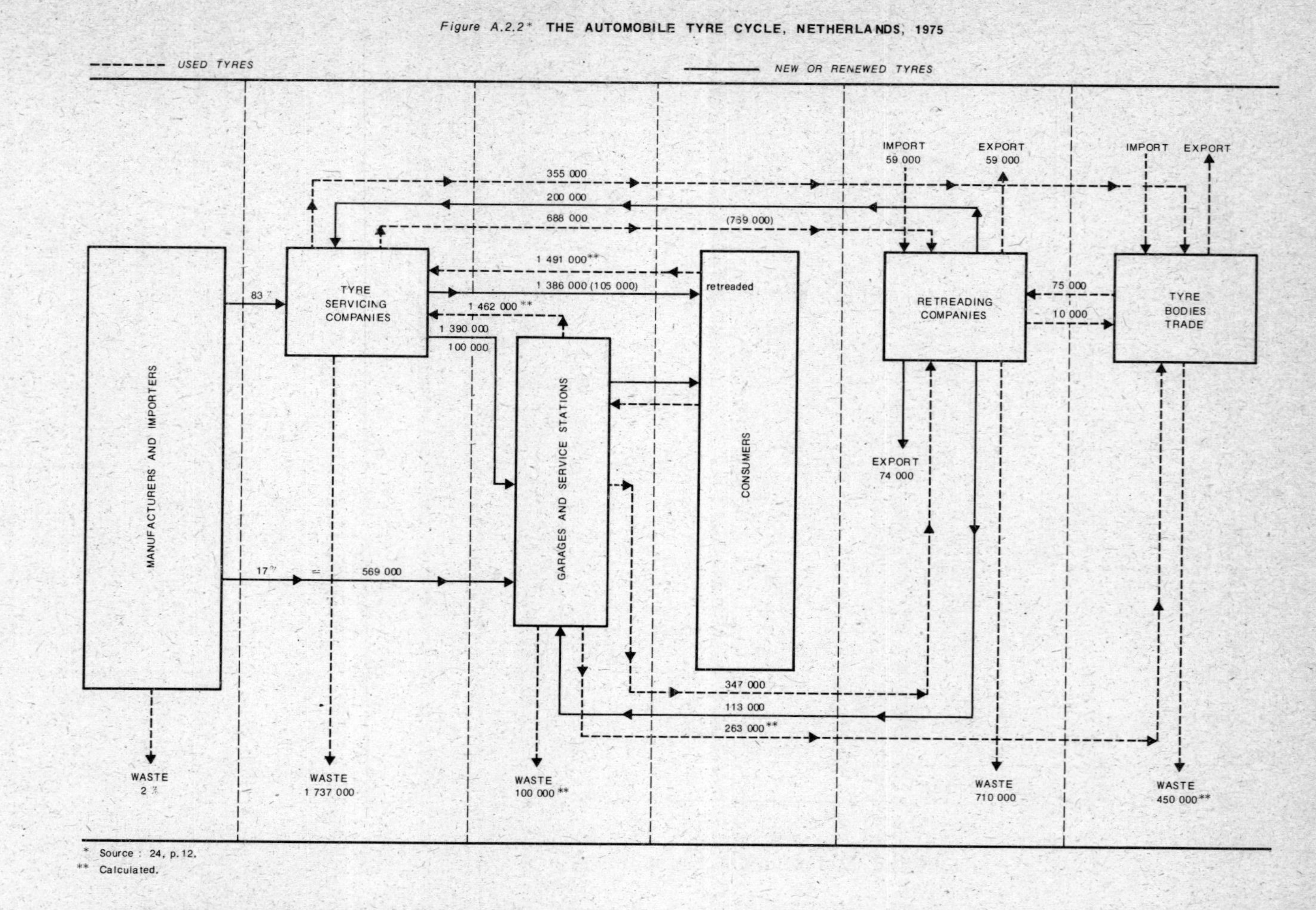

* Source : 24, p.12.

** Calculated.

*Figure A.2.3** **THE WASTE TYRE CYCLE IN THE GERMAN FEDERAL REPUBLIC**

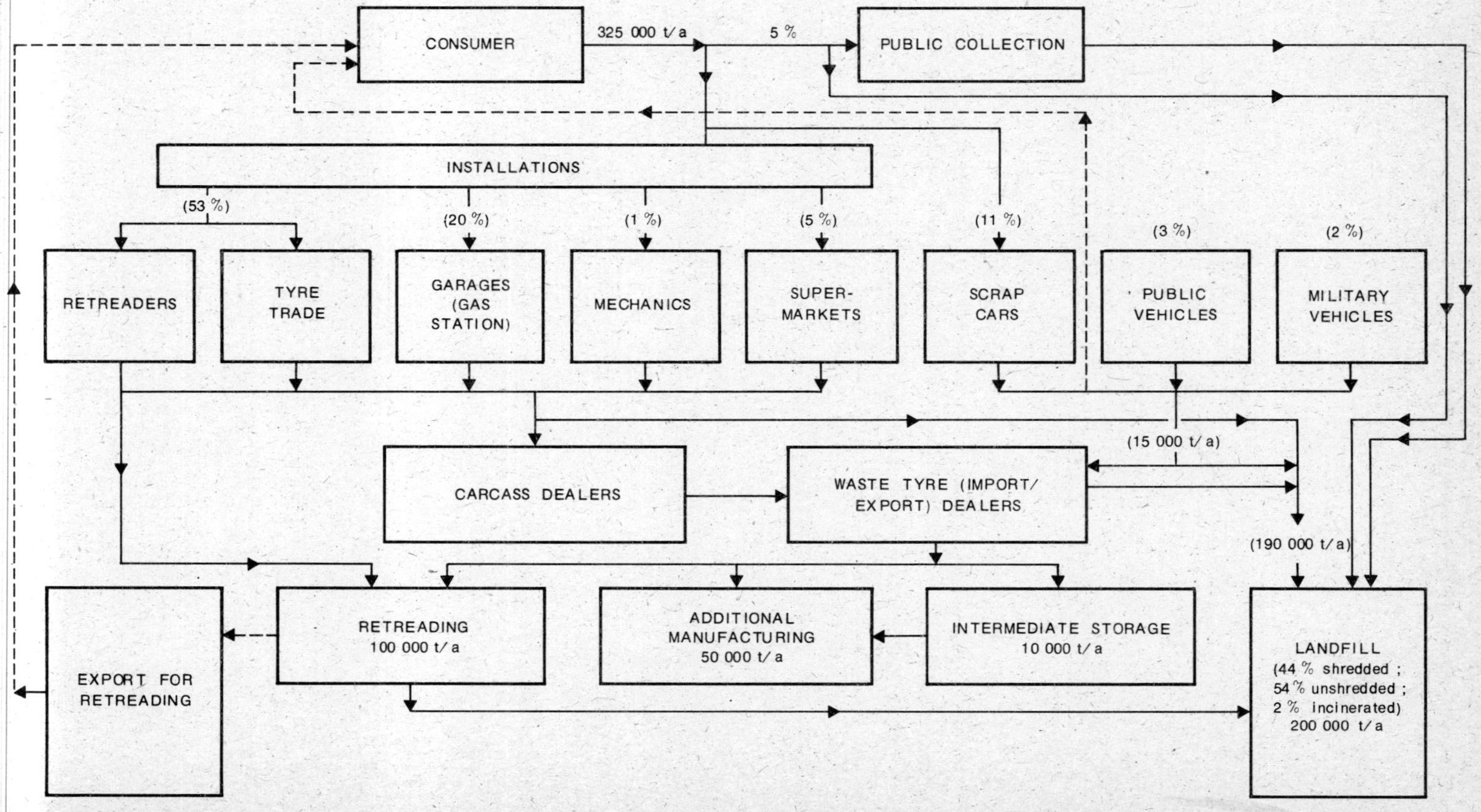

* Source : 2, p. IV/25.

Figure A.2.4* **THE FLOW OF TYRES IN THE TYRE TRADE**

United Kingdom

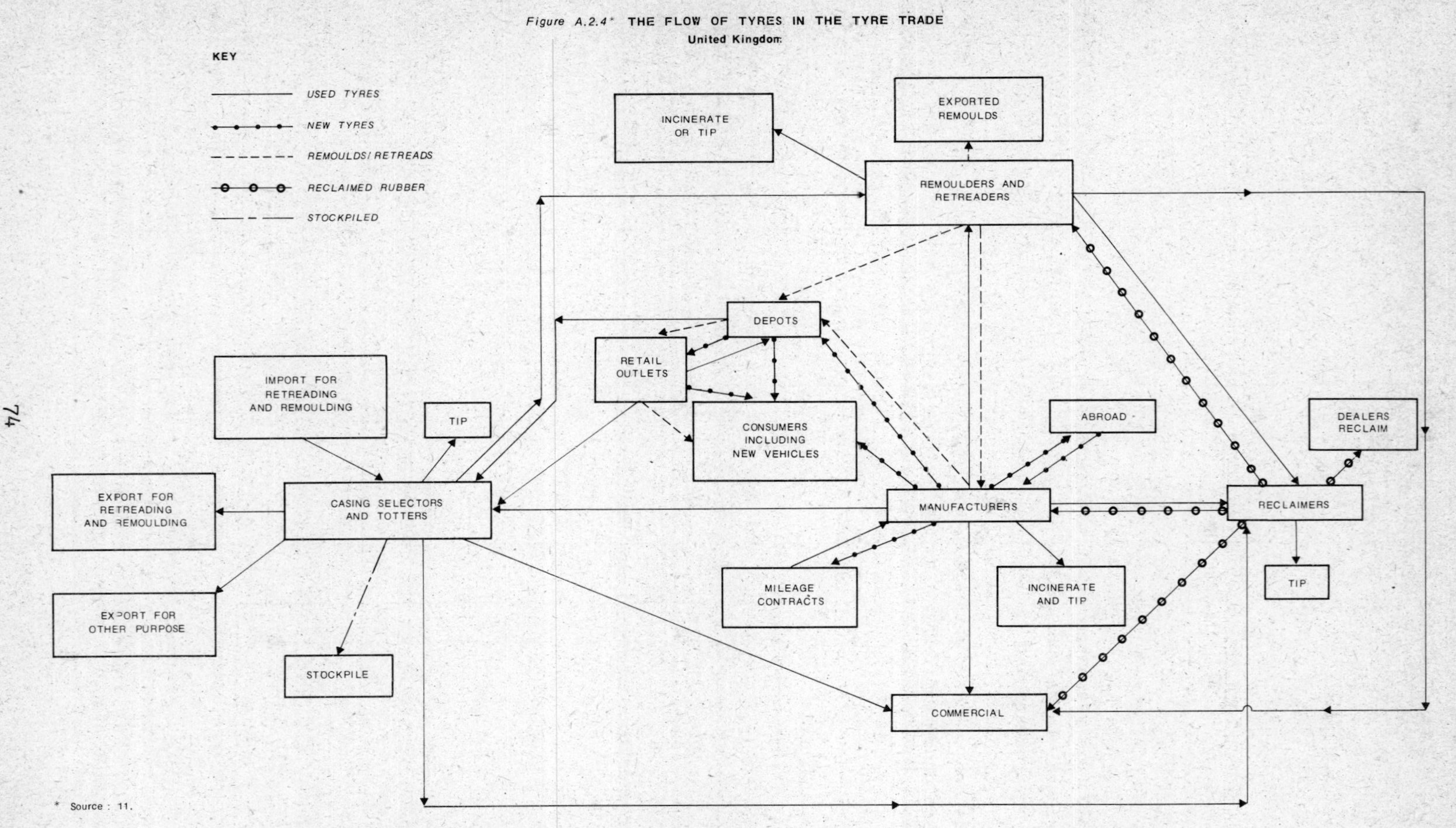

* Source : 11.

Annex 3

THE GENERATION AND DISPOSAL OF USED TYRES IN SELECTED MEMBER COUNTRIES

Figure A.3.1, based on the American tyre cycle, shows that, for 1969 (the year upon which the Figure is based), tyres going to ultimate disposal in the United States consisted of 13.6 million tyres rejected by retreaders and returned to retailers; 57.5 million tyres not examined by retreaders; and 7.2 million used truck tyres going from retail to disposal, these three making a total retail-disposal flow of 78.3 million used tyres. Tyres disposed of on junked vehicles amounted to 37.3 million, and 4 million were disposed of directly by consumers. Retreaders sent 64.3 million tyres for disposal. Thus, in 1969, the total flow of scrap tyres to ultimate disposal channels was approximately 183.9 million tyres.

Alternative estimates of the waste tyre flow in the United States are available, although the basis of these is not given. The volumes predicted by this alternate source differ sufficiently from those given above to merit their presentation. They are shown in Table A.3.1. The difference between the two sets of estimates is brought out more clearly when the figures are disaggregated into automobile discards and truck discards and truck and bus discards. These are shown in Table A.3.2.

Table A.3.1 *

WASTE GENERATION FROM TYRE DISCARD, 1972-1990

(millions)

Year	Quantity discarded
1972	122.5
1975	159.0
1980	184.5
1985	201.0
1990	223.0

* Source: 12, p. 178.

Figure A.3.1 * **TYRE MARKET MODEL**
United States, 1969

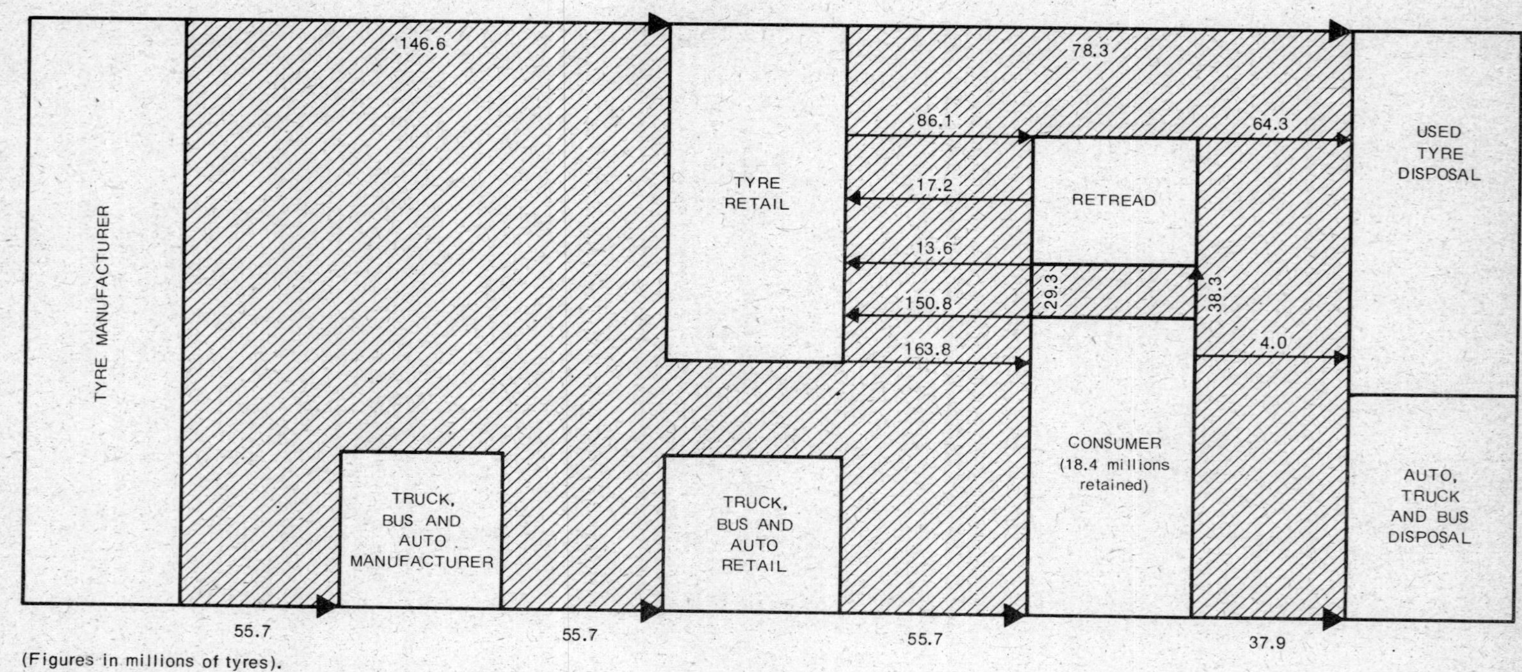

(Figures in millions of tyres).

* Showing the main flows of tyres from their manufacture to their disposal, and the quantities of tyres entering the flow at various stages.
Source : 9, p. 6 a.

Table A.3.2 *

WASTE GENERATION FROM PASSENGER CAR, AND TRUCK AND BUS TYRE DISCARD, UNITED STATES, 1972-1990

(millions)

Year	Passenger cars			Trucks and buses		
	Replacements	Retreads	Net discards	Replacements	Retreads	Net discards
1972	141	31	111	21	9.5	11.5
1975	175	31	144	25	10.0	15.0
1980	195	30	165	30	10.5	19.5
1985	215	30	180	32	11.0	21.0
1990	240	40	200	35	12.0	23.0

* Source: 12, p. 178.

Whilst Figure A.3.1 and Table A.3.2 are not directly comparable, it is possible to see that truck and bus tyres discarded appear to be higher in the latter, but that automobile tyre discards are considerably lower. There is no obvious reconciliation of these discrepancies.

In Sweden, the volume of waste tyres generated is estimated to be 70,000 tonnes per year. Of these about 5,000 tonnes go for the preparation of sports grounds and in the manufacture of blasting mats, fenders and similar products. About 18,000 tonnes are retreaded, but the largest proportion (about 47,000 tonnes, or 70 %) go to landfill.(1)

In the Federal Republic of Germany, the total volume of waste tyres generated is 340,000 tonnes, representing, by weight, 1 % of the total quantity of solid waste generated in this country. Of these about 60 % come from passenger cars, rather less than 30 % from trucks, and the remainder from NATO vehicles, construction and agricultural machinery, and similar relatively minor sources. The complete information is given in Table A.3.3. Typically, the largest proportion of these 340,000 tonnes of used tyres goes to landfill, either stredded into controlled landfill, or whole into uncontrolled landfill. Other major proportions are retreading, recycling and reclaim. The situation is shown in Table A.3.4.

Table A.3.3. *

WASTE TYRE GENERATION IN THE FEDERAL REPUBLIC OF GERMANY BY TYPE OF VEHICLE, 1974

(tonnes)

Vehicle	Waste tyres generated
Motorcycles	400
Cars	203,000
Station wagons	5,900
Buses	2,400
Lorries	104,200
Tracked vehicles	6,000
Agricultural vehicles	4,800
Other vehicles	13,300
Total	340,000

* Source: 2, p. IV/23.

Tables A.3.5, A.3.6 and A.3.7 show the volume of waste tyres generated in the United Kingdom, the latter two using the methodology discussed in chapter 3.

1) Information from personal communication.

Table A.3.4 *

DISPOSITION OF DISCARDED TYRES IN THE FEDERAL REPUBLIC OF GERMANY, 1974

Disposal route	Tonnes/Year (1)	Percentages
Reclaimed	20,000	6
Recycling (rubber crumb and buffings from retreading operations)	30,000	9
Pyrolysis	0	0
Incineration	7,000	2
Retreading	90,000	27
Export	0	0
Landfill: Shredded, controlled	85,000	25
Whole, uncontrolled	115,000	
Intermediate storage	100,000	30
Miscellaneous	6,000	1
	453,000	100

* Source: 2, p. IV/27.

1) The figures exceed 340,000 tonnes, because of an element of double counting: tyres going to other routes (retreading, reclaim) may ultimately be sent to the landfill sites.

In Denmark, the magnitude of the waste tyre problem has been studied recently [14]. The volume of waste tyres generated in Denmark between 1970 and 1975 is shown in Tables A.3.8 and A.3.9. Most discarded tyres go to landfill, others go to incineration, mechanical comminution, retreading, and the production of reclaimed rubber. Figures are not available for the proportion of used tyres going into each disposal route.

An enquiry into the disposal of used tyres in the Netherlands [24] showed that the total volume of tyres is approximately 4,700,000 tyres, or 43,770 tonnes of waste tyres. Table A.3.10 shows the situation in detail for 1973. As in most other countries, the majority of these discards go to landfill although predictably the proportion is much higher for car tyres than for truck and bus tyres. A higher proportion of truck and bus tyres go to incineration and to export, and a significantly higher proportion of truck and bus tyres are disposed of illegally. Table A.3.11 shows the detail, as assessed by a questionnaire which had a 72.6 % response rate in 1973.

In France, the situation concerning used tyres generation and disposal is shown in Tables A.3.12 and A.3.13, for the two years 1971 and 1976.

Table A.3.5 *

ESTIMATES OF UNITED KINGDOM ARISINGS OF SCRAP TYRES BY TYPE OF REINFORCEMENT, 1974/1980

Millions (%)

	Car/Van tyres			Truck/Bus tyres		
	1974	1977	1980	1974	1977	1980
Steel-belted radials	2.96 (15)	7.11 (33)	12.93 (55)	1.33 (54)	1.75 (70)	2.29 (85)
Textile-belted radials	5.92 (30)	7.55 (35)	7.05 (30)	-	-	-
Crossplies (textiles)	10.85 (55)	6.90 (32)	3.53 (15)	1.13 (46)	0.73 (30)	0.41 (15)
	19.73 (100)	21.56 (100)	23.51 (100)	2.46 (100)	2.50 (100)	2.70 (100)

* Source: 21, p. 14.

Table A.3.6 *

ESTIMATES OF UNITED KINGDOM ARISINGS OF SCRAP CAR/VAN TYRES - DETAILS OF CALCULATIONS

(all figures in thousand units)

	1970	1971	1972	1973	1974	1975	1976	1977	1978	1979	1980
1. Home replacement sales of new tyres (crossply + radial) (a)			16,426	15,371	14,385	14,283	14,663	15,000	15,344	15,716	16,110
2. Vehicles scrapped, fitted with car/van tyres			1,203	1,091	1,069	1,133	1,206	1,312	1,321	1,407	1,480
3. Tyres per vehicle			5	5	5	5	5	5	5	5	5
4. Tyres removed from scrapped vehicles (4) = (2) x (3)			6,015	5,455	5,345	5,665	6,030	6,560	6,605	7,035	7,400
5. Total scrap tyre arisings (5) = (1) + (4)			22,441	20,826	19,730	19,948	20,693	21,560	21,949	22,751	23,510
6. Radial percentage in new tyres (b)	37	45	51	62	68	74	80	85	88	92	94
7. Radial percentage in scrap tyres (6) lagged by 3 years				37	45	51	62	68	74	80	85
8. Radial scrap tyre arisings (b) (8) = (5) x (7) x 0.01				7,706	8,879	10,173	12,830	14,661	16,242	18,201	19,984
9. Crossply scrap tyre arisings (9) = (5) - (8)				13,120	10,851	9,775	7,863	6,889	5,707	4,550	3,526

a) Includes imports.

b) Original equipment and home replacement only.

* Source: 21, p. 13.

Table A.3.7 *

ESTIMATES OF UNITED KINGDOM ARISINGS OF SCRAP TRUCK/BUS TYRES - DETAILS OF CALCULATIONS

(all figures in thousand units)

	1970	1971	1972	1973	1974	1975	1976	1977	1978	1979	1980
1. Home replacement sales of new tyres (crossply + radial) (a)			1,965	1,611	1,825	1,748	1,791	1,853	1,904	1,951	2,009
2. Vehicles scrapped, fitted with truck tyres			93	74	79	75	77	81	83	85	86
3. Tyres per vehicle			8	8	8	8	8	8	8	8	8
4. Tyres removed from scrapped vehicles (4) = (2) x (3)			744	592	632	600	616	648	664	6..	688
5. Total scrap tyre arisings (5) = (1) + (4)			2,709	2,203	2,457	2,348	2,407	2,501	2,568	2.6..	2,697
6. Radial percentage in new tyres			54	60	64	70	74	79	85	87	90
7. Radial percentage in scrap tyres (6) lagged by 2 years (b)					54	60	64	70	74	79	85
8. Radial scrap tyre arisings (8) = (5) x (7) x 0.01					1,327	1,409	1,540	1,751	1,900	2,078	2,292
9. Crossply scrap tyre arisings (9) = (5) - (8)					1,130	939	867	750	668	553	405

a) Includes imports.

Original equipment and home replacement only.

b) TTRL use 9 as average. Recent observations shows 8 as being more realistic.

Table A.3.8 *

GENERATION OF WASTE TYRES FROM PASSENGER CARS, DENMARK, 1970-1975

('000's tyres)

Year	Import of new tyres	Correction for tyres on new cars	Trade in used tyres	Tyres from scrap cars	Retreads	Total discards
1970	1,693	71	+ 40	415	+ 15	2,002
1971	1,607	57	+ 32	410	+ 12	2,104
1972	1,578	57	+ 216	391	+ 22	2,150
1973	1,702	85	+ 224	518	+ 73	2,432
1974	1,092	57	+ 143	418	+ 6	1,602
1975	1,258	-	+ 350	442	- 24	2,026

* Source: 14, pp. 63-66.

Table A.3.9 *

GENERATION OF WASTE TYRES FROM TRUCKS DENMARK, 1970-1975

('000's tyres)

Year	Import of new tyres	Correction for tyres on new cars	Trade in used tyres	Tyres from scrap cars	Retreads	Total discards
1970	193	64	+ 75	41	-	175
1971	194	59	+ 15	45	+ 5	200
1972	138	52	+ 24	36	+ 9	155
1973	160	68	+ 24	42	+ 12	170
1974	147	42	+ 16	54	+ 1	186
1975	134	22	+ 39	50	+ 23	204

* Source: 14, pp. 63-66.

Table A.3.10 *

SOURCES OF WASTE TYRES, NETHERLANDS, 1973
(by numbers of tyres and by weight)

Source	Cars	Trucks and buses	
Tyre service	1,737,000	99,000	
Companies retreading	710,000	40,000	
Companies tyre body trade	403,000	30,000	
Garages	100,000	-	
Fleetowners	-	29,000	
Derelict vehicles	1,375,000	161,000	
Total	4,325,000	359,000	4,764,000
Weight	29,410 tonnes (at 6.8 kg/tyre)	14,360 tonnes (at 40 kg/tyre)	43,770 tonnes

* Source: 24 pp. 11-12.

Table A.3.11 *

DESTINATION OF WASTE TYRES, NETHERLANDS, 1973 (1)

(%)

Disposal route	Cars	Trucks and buses
Municipal cleansing departments	65.5	35.0
Incineration plants	5.2	15.6
Agricultural purposes	10.3	18.5
Export	5.0	14.5
Storage	8.6	3.0
Illegal waste disposal	3.9	12.0

1) Excluding tyres on derelict vehicles.

* Source: 24, p. 4.

Table A.3.12 *

QUANTITY OF USED TYRES IN FRANCE

Type	1971		1976		Percentage increase by weight
	millions of tyres	thousands of tonnes	millions of tyres	thousands of tonnes	
Passenger cars	18.3	126	20.4	143	+ 13 %
Vans and other vehicles	3.5	159	4.1	185	+ 16 %
Two-wheelers	10.6	7.8	6.2	4.5	- 42 %
Total	32.4	290	30.7	330	+ 13 %

* Source: 17, Annex III, Table 2.

Table A.3.13 *

DESTINATION OF WASTE TYRES IN FRANCE, 1971 AND 1976

(tonnes)

	1971	1976
Retreading		
- passenger vehicles	6,000	7,800
- vans	5,000	4,900
- heavy vehicles	43,700	58,700
Total retreaded	54,000	71,400
Reclaim	11,600	14,000
Rubber crumb	10,000	11,000
Total	76,700	96,400

* Source: 17, Table 3, p. 71.

Annex 4

STATISTICS ON RECLAIM RUBBER PRODUCTION, CONSUMPTIONS AND STOCKS

Table A.4.1 [1]

RECLAIMED RUBBER PRODUCTION, CONSUMPTION AND STOCKS

	United States			France	Germany, Fed. Rep.			United Kingdom			Australia			Brazil			Canada		
	Production	Consumption	Stocks	Production	Production	Consumption	Stocks	Production	Consumption	Stocks	Production	Consumption	Stocks	Production	Consumption	Stocks	Production	Consumption	Stocks
1967 ...	247,560	243,111	28,856	39,270	30,125	32,164	4,460	33,400	30,200	3,600	9,507	9,452	1,101	14,493	14,474	449	15,001	18,285	1,989
1968 ...	261,346	254,445	30,055	37,751	33,528	36,263	5,014	35,300	29,600	3,000	9,586	8,860	1,828	18,868	18,131	890	15,811	16,846	1,676
1969 ...	242,757	235,489	29,737	37,751	28,657	35,243	5,388	34,000	27,000	3,800	7,942	8,760	692	18,643	18,125	785	15,046	17,367	1,904
1970 ...	203,773	202,774	28,022	38,915	22,023	32,319	3,702	29,700	23,800	2,200	8,924	8,906	867	20,920	20,603	1,247	12,209	16,258	2,167
1971 ...	202,387	203,691	23,033	36,887	14,898	26,896	2,869	29,500	22,900	3,300	8,999	9,639	879	22,018	22,754	1,039	13,619	17,934	1,159
1972 ...	197,567	190,592	20,227	34,704	10,032	23,286	2,112	25,700	21,200	3,100	8,546	9,120	490	23,743	24,159	1,526	14,045	17,680	1,056
1973 ...	204,903	167,502	25,822r	32,648	13,725	20,255	2,455	25,900	19,900	1,900	9,310	10,893	320	27,830	27,471	2,114	8,010	18,104	1,476
1974 ...	135,885	139,672r	18,291r	33,218	14,546	18,394	2,172	25,800	17,000	2,300	9,180	9,640	520	29,448	29,383	3,418	7,655	13,332	2,353
1975 ...	120,578	119,665r	13,789r	33,316	10,843	16,444	1,607	24,600	17,800	2,000	5,810	6,030	460	28,536	28,460	2,269	3,878	12,467	1,610
1976 ...	110,701r	107,182r	17,117r	33,626p	14,149	15,737	1,369	24,300	16,400	2,200	4,237	5,336	224	31,577	31,818	2,341	4,623	11,212	1,104
1977 Year ...	121,098q	122,671q			13,308	14,217		23,000	16,700q		1,096	2,699		31,632	31,248		3,431	9,497	
1978 Jan. ...	9,446	9,785	14,756		1,000	1,150	1,750	2,000	1,400	2,000	50	155	110	2,839	2,823	2,509	244	712	1,237
Feb. ...	9,618	9,115	14,729		900	1,100	1,650	1,700	1,300	2,100	164	251	200	2,663	2,658	2,857	311	728	924
March ...	9,605	9,385	14,519		1,102q	1,250q	1,489	1,500	1,300	1,800	83	226	190	2,837	2,617	3,112	285	754	901
April ...	10,054	10,107	13,452		950	1,000	1,500	1,900	1,500	2,200	42	171	170	2,490	2,745	3,276	350	814	989
May ...	9,848	10,283	13,702		900	900	1,500	1,700	1,300	1,900	74	162	170	2,774	2,740	1,606	331	819	1,025
June ...	9,881	10,259	13,561		1,039q	1,179q	1,469	1,600	1,300	2,000	70	176	100	2,986	2,940	3,835	308	847	1,037
July ...	9,526	8,747	13,674		1,100	1,000	1,600	1,400	1,100	2,200				3,110	2,931	4,221	273	650	1,157
Aug. ...	10,789	9,599	15,139		900	800	1,450	1,300	1,200	2,000				2,894	2,816	4,300	194	678	988
Sept. ...					1,155q	1,054q	1,367							2,456	2,416	4,197	288	762	1,047
Oct.																			
Nov.																			
Dec.																			
Year ...																			

Source: Rubber Statistical Bulletin.

All figures in metric tonnes

.. = not available

o = less than half a tonne

r = revised

p = provisional

\- = nil

q ql = including quaterly or year-end adjustment

Annex 5

THE COSTS AND BENEFITS OF ORIGINAL TREAD VERSUS RETREADED TYRES IN THE UNITED KINGDOM, 1977

In 1972, the cost of accidents in 1977 prices and values was £24 million on motorways and £1,317 million on all purpose roads.

a) Following the 1973 analysis in assuming that:

i) 20 % of accidents on motorways were caused by tyre failure

ii) 26 % of the accidents on motorways caused by tyre failure were caused by the failure of retreads,

the cost of accidents caused by failure of retreads on motorways in 1972 was £1.25 million in 1977 values and prices. From sample data on the incidence of retreads in the tyre population (estimated to be 7 % in 1972) the failure rate of retreads <u>versus</u> original tread tyres on motorways can be calculated as 4.6:1. Therefore, if all retreads had been replaced by original tread tyres in 1972, the cost of £1.25 million would have been reduced to $\frac{£1.25}{4.6}$ million = £0.2713 million. The cost of accidents on motorways specifically attributable to the fact that some tyres were retreads in 1972 was, therefore,

£1.25 million - £0.2713 million = £0.98 million in 1977 values and prices.

b) Assuming that in 1972 :

i) 1 % of all accidents on all purpose roads were caused by tyre failure

ii) of these accidents, 18 % were due to the failure of retreads

and making a similar calculation to that in a) above, gives a relative failure rate of 3:1 for retreads relative to original tread tyres on all purpose roads, and an estimated cost of accidents specifically attributable to retread tyres of £1.58 million in 1977 values and prices.

In 1972, therefore, the total cost of accidents which could be specifically attributed to retreaded tyres was £2.56 million. Assuming that, in 1972, 7 % of all tyres in use were retreads, over a private vehicle population of about 13 million the average accident cost of retreads relative to original tread tyres is $\frac{£2.56 \text{ million}}{3.64}$ = about 70p per tyre per year. Again, following the 1973 cost:benefit analysis, the average life of a retread is taken to be between 2 and 3 years, giving a capital "accident cost" per tyre of between £1.35 and £1.96 in 1977 values and prices.

The Department of Trade and Industry supplied the following 1977 price information on retreads and original tread tyres:

(£)

	Price per tyre	
	New original tread tyres	Retreads
Radials:		
textile	22	13
steel	23	14
Cross-ply	14	7

There is no complete information available on the incidence of these various types of tyre in the tyre population, so it is not possible to make any precise calculation of the difference in price between an "average" original tread tyre and an "average" retread. However, it would seem reasonable to suppose that this cost difference is in the region of £6 to £7. If this estimate is fairly accurate and reflects a resource cost differential, then it would appear that the costs of replacing a retreaded tyre by an original tread tyre (c.£6) outweighs the safety benefits to be gained (£1.35 - £1.96) by a factor of at least 3:1. It can be concluded, therefore, that allowing for any change in the relative costs of accidents and tyres to 1977 does not change the conclusions of the original analysis.

In 1977 (in 1977 values and prices), the total cost of accidents on motorways was at least £28.7 million, and on other roads, at least £1,263.6 million. There is evidence to suggest that the proportion of retreads in the tyre population has decreased considerably over the last few years. If this is the case, then following from the estimated difference in risk of failure between original tread and retreads and assuming all other factors are constant, the percentage of total accidents due to tyre failure and to the failure of retread tyres would be expected to be lower in 1977 than in 1972. However, there is no data available on the incidence of tyre failure in accidents or any firm data on the proportion of retreads in the tyre population for 1977. In order to obtain an approximate estimate of the percentage of accidents due to tyre failure and due to retread failure in 1977, it is therefore necessary to adjust the 1972 estimates to an assumed lower incidence of retread tyres in the tyre population. If it can be assumed, for example, that in 1977, 3 1/2 per cent of the tyres in use were retreads then from the information above it could be expected that in 1977 18 % of the accidents on motorways and 0.95 % of accidents on other roads would be due to tyre failure. Of these accidents, 14 % and 10 % respectively might be due to the failure of retreaded tyres. This calculation does, of course, assume that on balance other factors affecting accidents have remained constant and that the risk of failure of retreads _versus_ original tread tyres has also remained unchanged.

Assuming that, in 1977, 18 % of accidents on motorways were due to tyre failure and that, of these, 14 % were due to the failure of retreads, the cost of accidents caused by the failure of retreaded tyres on motorways was:

$$£28.7 \text{ million} \times 0.18 \times 0.14 = £0.723 \text{ million in } 1977 \qquad (1)$$

Given a failure rate of retreads _versus_ original tread tyres of 4.6:1 on motorways, this cost could have been reduced to:

$$\frac{£0.723 \text{ million}}{4.6} = £0.157 \text{ million} \qquad (2)$$

if all retreads had been replaced by original tread tyres. Therefore, the costs of accidents on motorways specifically attributable to retreads was:

$$£0.723 \text{ million} - £0.157 \text{ million} = £0.566 \text{ million in } 1977. \qquad (3)$$

On the assumption that, in 1977, 0.95 % of accidents on all-purpose roads were caused by tyre failure and, of these, about 10 % were due to the failure of retreaded tyres, the cost of accidents caused by the failure of retreaded tyres, the cost of accidents caused by the failure of retreads on all purpose roads was:

$$£1{,}263.6 \text{ million} \times 0.0095 \times 0.1 = £1{,}200 \text{ million} \qquad (4)$$

in 1977. Given a failure rate of retreads versus original tyres of 3:1 on all purpose roads, the cost of accidents on these roads specifically attributable to retreads was:

$$£1.200 \text{ million} - £0.400 \text{ million} = £0.800 \text{ million in } 1977. \qquad (5)$$

In 1977, the total cost of accidents specifically attributable to retreads was (from equation (3) and equation (5)):

$$£0.566 \text{ million} - £0.800 \text{ million} = £1.366 \text{ million} \qquad (6)$$

in 1977 prices. The private vehicle population in the same year was about 14 million, representing 56 million tyres in use. If the proportion of retreads in the tyre population was 3.5 % in 1977, the average annual accident cost of retreaded tyres relative to original tyres was:

$$\frac{£1.366 \text{ million}}{1.96^{(1)} \text{ million}} = £0.697 \text{ per tyre}$$

1) 3.5 % of 56 million.

With an expected tyre life of two years, this represents a capital cost (1) of: £1.331 at a 10 % discount rate, and £1.359 at a 5 % discount rate. If expected tyre life is three years, the capital costs are £1.960 at 10 % discount rate, and £1.993 at a 5 % discount rate. The result is not particularly sensitive to assumptions about the percentage of retreaded tyres in the tyre population. If if is assumed that retread tyre use has remained constant since 1972, i.e. at about 7 % of the tyres in use, the accident cost per retreaded tyre per annum increases marginally.

Overall, therefore, although total accident costs in 1977 were lower than in 1972 (on a comparable basis), the relative increase in the cost of accidents on motorways tended to raise the costs attributable to retread tyres (from £2.56 million to £2.67 million on a comparable basis). However, between 1972 and 1977, the private vehicle stock increased more than in proportion to this increase in attributable costs, hence the cost per retread was lower in 1977 than in 1972.

It would appear that neither revaluing benefits and costs of replacement nor taking into account relative traffic growth on motorways since 1972 has any appreciable effect on the conclusions reached by the original analysis. Naturally, these adjustments cannot represent a complete up-dating but unless there is evidence of a significant increase in the relative accident risk of retreads, it seems reasonable to expect that the costs of replacing retreads by original tread tyres would outweight the safety benefits to be gained by a factor of at least 3. Any trend towards safer retreads would, of course, reinforce this conclusion.

---

1) That is, a capitalised value of the accident costs over the 2 year life of the tyre.

Annex 6

TYRES DISPOSAL PROGRAMME IN PORTLAND, OREGON

The Metropolitan Service District of Portland, Oregon, has developed, since 1973, a comprehensive scrap tyre disposal programme. [20] The programme is designed to prevent the illicit dumping of scrap tyres in the metropolitan area of Portland. The Metropolitan Service District carries out the following functions:

i) licences certain landfill sites to accept tyres from the public (not more than 30 tyres per person per day) for on-site disposal, which are shredding before burying

ii) issues permits to tyre carriers (excluding low volume carriers) for a nominal fee, and accepts a $1,000 bond from the carrier

iii) requires registered carriers to deliver all scrap tyres to a licensed disposal site, a registered tyre processing centre, or a registered tyre salvage centre

iv) requires anyone generating scrap tyres to give them to a registered carrier, return the tyres to the owner, or deliver them himself to a disposal site

v) licenses tyre disposal centres (license fee $10,000) for acceptance of discarded tyres, the centres being entitled to charge an acceptance fee of between 25-85 cents

vi) licenses tyre salvage centres, which must salvage all tyres they accept.

The programme is financed by a surcharge of 3 cents per tyre paid by processing and salvage centres, and on tyres taken out of the district for salvage by tyre carriers. Other income comes from licensing tyre carriers. It is estimated that, due to high legal costs, these fees have covered 50 % of the administration costs of the programme.

Annex 7

BIBLIOGRAPHY

1. (Anon): "Energy Recovery and the Problem of Tyre Disposal in Connecticut", paper given to the National Tyre Disposal Symposium, Washington, D.C., June 1977.

2. Bundesministerium des Innern: Materialien 2/76:IV Altreifen, Materialien zum Abfallwirtschafts Programme '75, Bonn, 1975.

3. Candle, R.D., and R.E. Payne: Cost Estimates to Furnish and Install the Proposed Modular-Mat-type Scrap Tyre Floating Breakwater, Goodyear Tyre and Rubber Co., Akron, Ohio, November 1974.

4. Chiesa, A., and G. Ghilardi: Evaluation of Tyre Abrasions in Terms of Driving Severity, Automatove Engineering Congress and Exposition, Detroit, Michigan, February 1975.

5. Continental Rubber B.V.: Onderzoekrapport uit de Period Januari t/m April, Amersfoort, April 1975, Mimeo.

6. Department of Environment, Housing and Community Development, Research Directorate: Tyre Recycling, Re-use and Disposal, Australian Government Publishing Service, Canberra, 1977.

7. The Evening Bulletin, Washington, D.C. February 13th, 1978.

8. Gobhall, W.W.: "A Process for the Manufacture of Carbon Black from Scrap Rubber", paper given to the National Tyre Disposal Symposium, Washington, D.C., 1977.

9. International Research and Technology Corporation: Tyre Recycling and Re-use Alternatives, Washington, D.C., 1974.

10. Janning, J., W. Kaminsky and H. Sinn: "Pyrolysis of Plastic Waste and Scrap Tyres in a Fluid Bed Reactor", United Nations Economic Commission for Europe, Seminar on Recycling of High Polymer Wastes, Dresden, German Democratic Republic, September 1978.

11. Local Government Operational Research Unit: Used Tyre Disposal in Britain, Royal Institute of Public Administration, Report C177, Manchester, September 1973.

12. Midwest Research Institute: Base Line Forecasts of Resource Recovery, 1972-1990, Kansas City, March 1975.

13. The National Tyre Dealers and Retreaders Association, Inc. Retreading Institute. National Tyre Disposal Symposium: Papers and Proceedings, Washington, D.C., 1977.

14. Office of Environmental Protection: Waste Tyres: Disposal, Re-use, Recovery, Copenhagen, June 1977 (Draft Report).

15. Office of Research and Development, Federal Highway Administration. Rubber-Asphalt Binder for Seal Coat Construction: Implementation Package, U.S. Department of Transportation, Washington, D.C. 73-1, 1973.

16. C. Peterson: Statement to the National Tyre Disposal Symposium, Washington, D.C., June 1977.

17. Rapport de la Commission, Organisation et financement au Comité national pour la récupération et l'élimination des déchets: Annexe III: Organisation et financement de la récupération et l'élimination des déchets de pneumatiques, Paris, June, 1978.

18. Report on Economic Analysis of Resources Recycling Analysis of Collecting Structure - Waste tyres Collected Dust and Waste Refractory, Clean Japan Centre, Recycling Series No. 3, 78-3, Tokyo, January 1979.

19. Research Directorate: Technical and Economic Survey of Rubber Waste Recovery in the EEC, Commission of the European Communities, Brussels, 1977.

20. Rhoten, C: "Scrap Tyre Processing and Disposal Programmes", paper given to the National Tyre Disposal Symposium, Washington, D.C., June 1977.

21. Rubber and Plastics Research Association: A Study of the Reclamation and Re-use of Waste Tyres, Shrewsbury, England, 1976.

22. Sandford, G., P. van Schaik and K.T. Solomon: Determining Cubic Freight Transportation Costs, Australian Road Research Board Report No. 34, December 1974.

23. Stevenson and Kellog Ltd.: Major Domestic Appliances and Automobile Tyres: Environmental and Economic Impacts of Product Durability, Toronto, 1977.

24. Stichting Verwijdering Afvalstoffen: Afgedanka Autobanden, Amersfoort, Netherlands, May 1976.

25. United States Environmental Protection Agency: Incentives for Tyre Recycling and Re-use, Washington, D.C., 1974.

26. Urban Systems Research and Engineering Inc.: Status, Trends and Impediments to Discarded Tyre Collection and Resource Recovery (Working Papers for May 1st Meeting), U.S.E.P.A. contract number 68-03-2725, Cambridge, Massachusetts, April 1979.

27. W.L. Wardropp and Associates, Ltd.: A Study of Waste Rubber Utilisation in the Prairie Provinces of Canada, Environmental Impact Control Directorate, Waste Management Branch Report EPS 3-EC-77-15, Fisheries and Environment Canada, Ottawa, August 1977.

28. R.R. Westerman: Tyres: Decreasing Solid Wastes and Manufacturing Throughput: Markets, Profits and Resource Recovery, EPA-600/5-78-009, Municipal Environmental Research Laboratory, Cincinnati, Ohio, July 1978.

29. Wilcox, J. "Municipal Waste: Economic Aspects of Technological Alternatives", Chapter 6 in D.W. Pearce and I. Walter (eds.): Resource Conservation, New York University Press, 1977.

30. Yakowitz, H. "The Interaction of PL94-580 with Uses Discarded Tyres", paper presented to the National Tyre Disposal Symposium, Washington, D.C., June 1977.

OECD SALES AGENTS
DÉPOSITAIRES DES PUBLICATIONS DE L'OCDE

ARGENTINA – ARGENTINE
Carlos Hirsch S.R.L., Florida 165, 4° Piso (Galería Guemes)
1333 BUENOS AIRES, Tel. 33.1787.2391 y 30.7122

AUSTRALIA – AUSTRALIE
Australia & New Zealand Book Company Pty Ltd.,
23 Cross Street, (P.O.B. 459)
BROOKVALE NSW 2100. Tel. 938.2244

AUSTRIA – AUTRICHE
OECD Publications and Information Center
4 Simrockstrasse 5300 BONN. Tel. (0228) 21.60.45
Local Agent/Agent local :
Gerold and Co., Graben 31, WIEN 1. Tel. 52.22.35

BELGIUM – BELGIQUE
LCLS
44 rue Otlet, B 1070 BRUXELLES. Tel. 02.521.28.13

BRAZIL – BRÉSIL
Mestre Jou S.A., Rua Guaipa 518,
Caixa Postal 24090, 05089 SAO PAULO 10. Tel. 261.1920
Rua Senador Dantas 19 s/205-6, RIO DE JANEIRO GB.
Tel. 232.07.32

CANADA
Renouf Publishing Company Limited,
2182 St. Catherine Street West,
MONTRÉAL, Quebec H3H 1M7. Tel. (514)937.3519
522 West Hasting,
VANCOUVER, B.C. V6B 1L6. Tel. (604) 687.3320

DENMARK – DANEMARK
Munksgaard Export and Subscription Service
35, Nørre Søgade
DK 1370 KØBENHAVN K. Tel. +45.1.12.85.70

FINLAND – FINLANDE
Akateeminen Kirjakauppa
Keskuskatu 1, 00100 HELSINKI 10. Tel. 65.11.22

FRANCE
Bureau des Publications de l'OCDE,
2 rue André-Pascal, 75775 PARIS CEDEX 16. Tel. (1) 524.81.67
Principal correspondant :
13602 AIX-EN-PROVENCE : Librairie de l'Université.
Tel. 26.18.08

GERMANY – ALLEMAGNE
OECD Publications and Information Center
4 Simrockstrasse 5300 BONN Tel. (0228) 21.60.45

GREECE – GRÈCE
Librairie Kauffmann, 28 rue du Stade,
ATHÈNES 132. Tel. 322.21.60

HONG-KONG
Government Information Services,
Sales and Publications Office, Baskerville House, 2nd floor,
13 Duddell Street, Central. Tel. 5.214375

ICELAND – ISLANDE
Snaebjörn Jönsson and Co., h.f.,
Hafnarstraeti 4 and 9, P.O.B. 1131, REYKJAVIK.
Tel. 13133/14281/11936

INDIA – INDE
Oxford Book and Stationery Co. :
NEW DELHI, Scindia House. Tel. 45896
CALCUTTA, 17 Park Street. Tel. 240832

INDONESIA – INDONÉSIE
PDIN-LIPI, P.O. Box 3065/JKT., JAKARTA, Tel. 583467

IRELAND – IRLANDE
TDC Publishers – Library Suppliers
12 North Frederick Street, DUBLIN 1 Tel. 744835-749677

ITALY – ITALIE
Libreria Commissionaria Sansoni :
Via Lamarmora 45, 50121 FIRENZE. Tel. 579751
Via Bartolini 29, 20155 MILANO. Tel. 365083
Sub-depositari :
Editrice e Libreria Herder,
Piazza Montecitorio 120, 00 186 ROMA. Tel. 6794628
Libreria Hoepli, Via Hoepli 5, 20121 MILANO. Tel. 865446
Libreria Lattes, Via Garibaldi 3, 10122 TORINO. Tel. 519274
La diffusione delle edizioni OCSE è inoltre assicurata dalle migliori librerie nelle città più importanti.

JAPAN – JAPON
OECD Publications and Information Center,
Landic Akasaka Bldg., 2-3-4 Akasaka,
Minato-ku, TOKYO 107 Tel. 586.2016

KOREA – CORÉE
Pan Korea Book Corporation,
P.O. Box n° 101 Kwangwhamun, SÉOUL. Tel. 72.7369

LEBANON – LIBAN
Documenta Scientifica/Redico,
Edison Building, Bliss Street, P.O. Box 5641, BEIRUT.
Tel. 354429 – 344425

MALAYSIA – MALAISIE
and/et **SINGAPORE - SINGAPOUR**
University of Malaysia Co-operative Bookshop Ltd.
P.O. Box 1127, Jalan Pantai Baru
KUALA LUMPUR. Tel. 51425, 54058, 54361

THE NETHERLANDS – PAYS-BAS
Staatsuitgeverij
Verzendboekhandel Chr. Plantijnnstraat
S-GRAVENAGE. Tel. nr. 070.789911
Voor bestellingen: Tel. 070.789208

NEW ZEALAND – NOUVELLE-ZÉLANDE
Publications Section,
Government Printing Office,
WELLINGTON: Walter Street. Tel. 847.679
Mulgrave Street, Private Bag. Tel. 737.320
World Trade Building, Cubacade, Cuba Street. Tel. 849.572
AUCKLAND: Hannaford Burton Building,
Rutland Street, Private Bag. Tel. 32.919
CHRISTCHURCH: 159 Hereford Street, Private Bag. Tel. 797.142
HAMILTON: Alexandra Street, P.O. Box 857. Tel. 80.103
DUNEDIN: T & G Building, Princes Street, P.O. Box 1104.
Tel. 778.294

NORWAY – NORVÈGE
J.G. TANUM A/S Karl Johansgate 43
P.O. Box 1177 Sentrum OSLO 1. Tel. (02) 80.12.60

PAKISTAN
Mirza Book Agency, 65 Shahrah Quaid-E-Azam, LAHORE 3.
Tel. 66839

PHILIPPINES
National Book Store, Inc.
Library Services Division, P.O. Box 1934, MANILA.
Tel. Nos. 49.43.06 to 09, 40.53.45, 49.45.12

PORTUGAL
Livraria Portugal, Rua do Carmo 70-74,
1117 LISBOA CODEX. Tel. 360582/3

SPAIN – ESPAGNE
Mundi-Prensa Libros, S.A.
Castello 37, Apartado 1223, MADRID-1. Tel. 275.46.55
Libreria Bastinos, Pelayo 52, BARCELONA 1. Tel. 222.06.00

SWEDEN – SUÈDE
AB CE Fritzes Kungl Hovbokhandel,
Box 16 356, S 103 27 STH, Regeringsgatan 12,
DS STOCKHOLM. Tel. 08/23.89.00

SWITZERLAND – SUISSE
OECD Publications and Information Center
4 Simrockstrasse 5300 BONN. Tel. (0228) 21.60.45
Local Agents/Agents locaux
Librairie Payot, 6 rue Grenus, 1211 GENÈVE 11. Tel. 022.31.89.50
Freihofer A.G., Weinbergstr. 109, CH-8006 ZÜRICH.
Tel. 01.3624282

TAIWAN – FORMOSE
National Book Company,
84-5 Sing Sung South Rd, Sec. 3, TAIPEI 107. Tel. 321.0698

THAILAND – THAILANDE
Suksit Siam Co., Ltd., 1715 Rama IV Rd,
Samyan, BANGKOK 5. Tel. 2511630

UNITED KINGDOM – ROYAUME-UNI
H.M. Stationery Office, P.O.B. 569,
LONDON SE1 9NH. Tel. 01.928.6977, Ext. 410 or
49 High Holborn, LONDON WC1V 6 HB (personal callers)
Branches at: EDINBURGH, BIRMINGHAM, BRISTOL,
MANCHESTER, CARDIFF, BELFAST.

UNITED STATES OF AMERICA – ÉTATS-UNIS
OECD Publications and Information Center, Suite 1207,
1750 Pennsylvania Ave., N.W. WASHINGTON D.C.20006.
Tel. (202) 724.1857

VENEZUELA
Libreria del Este, Avda. F. Miranda 52, Edificio Galipan,
CARACAS 106. Tel. 32.23.01/33.26.04/33.24.73

YUGOSLAVIA – YOUGOSLAVIE
Jugoslovenska Knjiga, Terazije 27, P.O.B. 36, BEOGRAD.
Tel. 621.992

Les commandes provenant de pays où l'OCDE n'a pas encore désigné de dépositaire peuvent être adressées à :
OCDE, Bureau des Publications, 2, rue André-Pascal, 75775 PARIS CEDEX 16.

Orders and inquiries from countries where sales agents have not yet been appointed may be sent to:
OECD, Publications Office, 2 rue André-Pascal, 75775 PARIS CEDEX 16.

63518-12-1980

OECD PUBLICATIONS, 2 rue André-Pascal, 75775 Paris Cedex 16 - No. 41 671 1980
PRINTED IN FRANCE
(600 DQ 97 80 07 1) ISBN 92-64-12131-5